ALLE ZEIT WACH 1842

P. Strata E. Carbone (Eds.)

Mg^{2+} and Excitable Membranes

With 62 Figures

Springer-Verlag
Berlin Heidelberg New York
London Paris Tokyo
Hong Kong Barcelona

Professor Piergiorgio Strata
Professor Emilio Carbone
University of Turin
Department of Human Anatomy and Physiology
Corso Raffaello 30
I-10125 Turin, Italy

Library of Congress Cataloging-in-Publication Data
Mg^{2+} and excitable membranes / edited by Piergiorgio Strata and Emilio Carbone. p. cm. Based on a symposium held in September 1989 on the occasion of the 12th Annual Meeting of the European Neuroscience Association in Torino. On t.p. "2^{+}" is superscript. Includes bibliographical references. Includes index.

ISBN-13: 978-3-642-75638-2 e-ISBN-13: 978-3-642-75636-8
DOI: 10.1007/978-3-642-75636-8

1. Magnesium–Physiological effect–Congresses. 2. Excitable membranes–Effect of metals on–Congresses. I. Strata, Piergiorgio, 1935– . II. Carbone, Emilio, 1948– . III. European Neuroscience Congress (12th : 1989 : Turin, Italy) IV. Title: Magnesium two plus and excitable membranes. [DNLM: 1. Cell Membrane–drug effects–congresses. 2. Cell Membrane–physiology–congresses. 3. Ion Channels–physiology–congresses. 4. Magnesium–pharmacology–congresses. QH 601 M196 1989] QP535.M4M42 1991 591.87′5–dc20 DNLM/DLC for Library of Congress 90-10405 CIP

Softcover reprint of the hardcover 1st edition 1991

25/3145-543210 – Printed on acid-free paper

Preface

The present book is the outcome of a symposium on magnesium and excitable membranes, held in September 1989 on the occasion of the 12th annual meeting of the European Neuroscience Association in Torino. Leading specialists from around the world met in order to discuss the physiological and clinical effects of magnesium on excitable cells.

Magnesium ions are known to play a key role in the regulation of a variety of cell functions and, as such, have been implicated in the control of neuronal excitability and cardiovascular diseases. In the last decade, the advent of new electrophysiological techniques has provided important details on the modulatory action of magnesium on ion channels and the receptor proteins of neurotransmitters. Thus, in neuronal and muscle cells, extracellular and intracellular magnesium is shown to block ion passage through open channels in a voltage- and time-dependent manner. The mechanism appears to be common for voltage- and receptor-operated channels, and it is thought to be an important cofactor in the long-term potentiation of synaptic transmission, brain aging, and cardiac activity. Recent findings show that the action of internal magnesium has additional features. High and low affinity sites for internal magnesium are involved in the regulation of channel gatings and ion permeation. Biochemical studies suggest that internal magnesium interferes with G-protein subunits, GTP-ase activity, and second messengers involved in channel phosphorylation. Considerable progress has also been made in the use of magnesium to attenuate or prevent neurotrauma and in understanding its role in brain energy metabolism.

This book contains reports of some of the most prominent work being done in this field. It also includes a short review on the properties and structural diversities of membrane ion channels and an introductory chapter on the methods of detecting intracellular free magnesium using ion selective microelectrodes.

We would like to thank Lirca Synthelabo for supporting the organization of the symposium and the publication of this volume. We also express our acknowledgements to Mrs. J. Hamilton and the staff of Springer-Verlag for their excellent collaboration and to

Miss G. Milano for her help at all stages of the organization and her secretarial assistance.

Turin, February 1990

P. Strata
E. Carbone

Contents

Use of Ion Selective Microelectrodes to Measure Intracellular Free Mg^{2+}
John A. S. McGuigan, Lothar A. Blatter, and Arlette Buri 1

Voltage-Gated Ionic Channels: Diversity and Modulation by Mg^{2+}
Arthur M. Brown and Atsuko Yatani 21

ATP-Dependent K^+ Channels of Excitable Membranes
Peter R. Stanfield 33

Voltage-Dependent Blockage of Cardiac Inwardly Rectifying K^+ Channels by Internal Mg^{2+}
Hiroko Matsuda 51

Regulation of the Voltage-Gated Ca^{2+} Current by Intracellular Free Mg^{2+} Studied by Internal Perfusion of Single Cardiac Myocytes
H. Criss Hartzell and Richard E. White 71

Blockage of Neuronal Low-Threshold Ca^{2+} Channels by Extracellular Mg^{2+}
H. Dieter Lux and Emilio Carbone 97

Dual Blockage by Mg^{2+} of the *N*-Methyl-D-Aspartate-Activated Channel
John W. Johnson and Philippe Ascher 105

Excess Mg^{2+} and Central Nervous System Metabolism
Gregory Crosby and Michele D. Szabo 119

Mg^{2+} in Neurotrauma: Its Role and Therapeutic Implications
Robert Vink, Tracy K. McIntosh, and Alan I. Faden . . 125

Subject Index 147

Contributors

You will find the addresses at the beginning of the respective contributions

Ascher, Philippe 105
Blatter, Lothar A. 1
Brown, Arthur M. 21
Buri, Arlette 1
Carbone, Emilio 97
Crosby, Gregory 119
Faden, Alan I. 125
Hartzell, H. Criss 71
Johnson, John W. 105
Lux, H. Dieter 97
Matsuda, Hiroko 51
McGuigan, John A. S. 1
McIntosh, Tracy K. 125
Stanfield, Peter R. 33
Szabo, Michele D. 119
Vink, Robert 125
White, Richard E. 71
Yatani, Atsuko 21

Use of Ion Selective Microelectrodes to Measure Intracellular Free Mg^{2+}

John A.S. McGuigan[1], Lothar A. Blatter[1,2] and Arlette Buri[1]

Mg^{2+} has been dubbed "the forgotten cation" (Brady et al. 1987) for although it is the second most common intracellular cation, the study of its homeostasis and its regulatory function in health and disease has long been neglected. Gradually, however, it is becoming clear that Mg^{2+} plays an important intracellular role: It is a co-factor in numerous enzymatic reactions, has a regulatory function on several ionic channels, influences Ca^{2+} uptake into the mitochondria and K^+ uptake by skeletal muscle, and may play a crucial role in cardiovascular disease (see reviews by Altura 1988; Elin 1988; Shils 1988; Lauter 1989; White and Hartzell 1989; Review in Am J Med 1987).

Up to now the major problem in the understanding of the role of Mg^{2+} in cellular processes has been the difficulty in measuring the free concentration of the ion ($[Mg^{2+}]_i$) for it is this free concentration not the total concentration that is the physiological parameter. In heart muscle the total concentration of magnesium has been found to be 17.3 m*M* (Page and Polimeni 1972), but the free concentration is around 1 m*M* (Blatter and McGuigan 1988). This means that due to the large buffer capacity for magnesium in the cell, a decrease in total magnesium could lead to an increase in $[Mg^{2+}]_i$ (Corkey et al. 1986); hence, the importance of estimating the free intracellular concentration of the ion.

The most direct method of measuring the intracellular free magnesium concentration is to use ion selective microelectrodes. However, the magnesium sensor ETH 1117 reacts intracellularly not only to Mg^{2+} but also to Na^+ and K^+, which limits its usefulness (Blatter and McGuigan 1988). Because of these disadvantages, Simon's group at the ETH in Zürich have recently produced a new magnesium resin with much better selectivity against Na^+ and K^+ than ETH 1117 (Hu et al. 1989). Two other new methods for Mg^{2+} measurement have been introduced, namely a fluorinated NMR method (Kirschenlohr et al. 1988; Levy et al. 1988) and a fluorescent intracellular probe (Raju et al. 1989). While these new methods have their advantages, this article will be limited to a critical look at the use of ion selective microelectrodes to measure intracellular magnesium. General references for the use of ion selective microelectrodes are Thomas (1978) and Ammann (1986).

---

[1] Institute of Physiology, Bühlplatz 5, CH-3012 Berne, Switzerland
[2] Present address: Department of Physiology, University of Maryland, School of Medicine, Baltimore, MD 21201, USA

Microelectrodes

Ion selective microelectrodes are glass microelectrodes, capable of being inserted into small cells, whose tips are filled with a sensor which is highly specific for an individual ion, in this case for Mg^{2+}. The resins are hydrophobic, so the electrodes must first be silanized, then filled with the magnesium sensor ETH 1117 (Lanter et al. 1980) or with the new sensor ETH 5214 (Hu et al. 1989). They are then backfilled with 100 m*M* $MgCl_2$.

Calibration is carried out against a 3 *M* KCl reference electrode in solutions mimicking the intracellular ionic composition but with varying Mg^{2+} concentrations (Fig. 1 A). The potential in Tyrode's solution is defined as zero, and the 3 *M* reference electrode minimizes changes in junction potentials.

From the measurements in the calibration chamber a calibration curve can be constructed (Fig. 1 B), and this allows a measured magnesium potential U_{Mg} to be converted into a concentration. Actual measurement is complicated by the fact that muscle cells possess a membrane potential E_m, and this is also measured by the ion selective microelectrode (Fig. 1 C). E_m must be measured independently and subtracted from the potential of the ion selective microelectrode (V_{Mg}) to obtain the magnesium signal U_{Mg}. Heart cells form a synctium, so in a small bundle at rest it is a fair assumption that the membrane potential is uniform throughout. This has the advantage that the ion selective electrode and the membrane potential electrode need not be in the same cell. Where a synctium is not formed, then two electrodes must be inserted into the same cell. Although it can be assumed that under physiological conditions in heart muscle the membrane potential is uniformly distributed, this is not always so under pathological conditions (cf. Chapman and Suleman 1986). One way round this problem is to use double-barrelled microelectrodes.

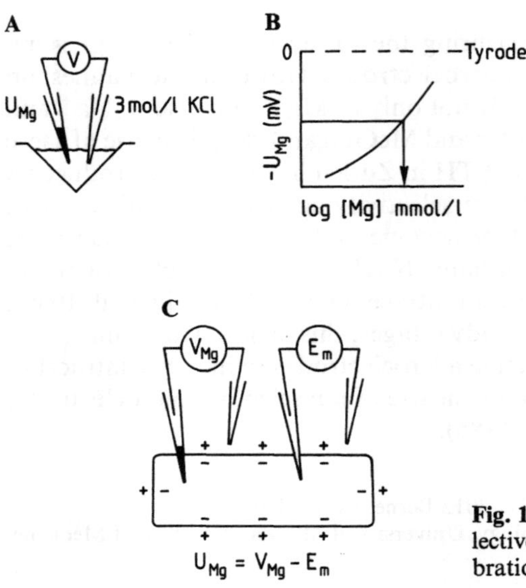

Fig. 1 A–C. Principles of the use of ion selective microelectrodes: **A** calibration, **B** calibration curve and **C** measurement in a cell

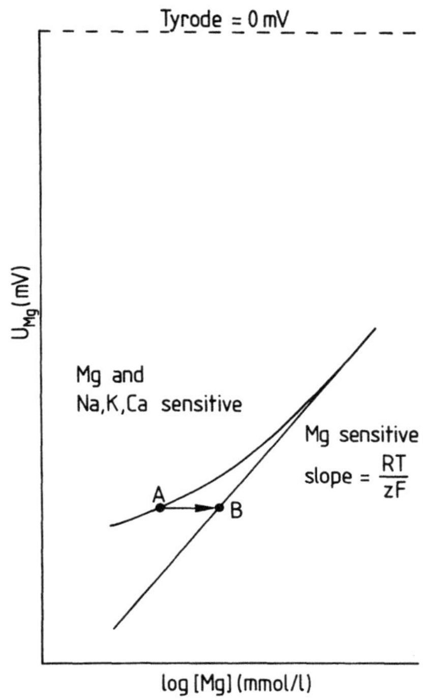

Fig. 2. Comparison of the response of a Mg^{2+} selective microelectrode that only reacts to magnesium and one that reacts to other ions as well

Selectivity

If the electrodes only reacted to Mg^{2+} the calibration curve would be a straight line. However, as is shown in Fig. 2, a deviation from linearity occurs if the resin reacts to other ions, e.g. Ca^{2+}, Na^{+}, and K^{+}. This arises because the electrodes, although to a greatly diminished extent, treat these interfering ions as if they were Mg^{2+} ions and read the "higher" Mg^{2+} concentration B instead of the true concentration A. The behaviour of the Mg^{2+} electrodes in the presence of interfering ions can be described by the semi-empirical Nicolsky-Eisenman equation (Ammann 1986)

$$U = U_0 + (R^* T/F)^* \ln [a_{Mg} + \sum K_i^* (a_i)^{Z_{Mg}/Z_i}]$$

where R, T, and F have their usual significance, and z is the valency. U is the measured potential, U_0 is a constant potential of the measuring system. K is a dimensionless selectivity coefficient which, if the electrode was ideally selective, would be zero.

In this equation, a is the single ion activity for Mg^{2+} (a_{Mg}) and the interfering ions (a_i). The activity is calculated by multiplying the concentration of the ion by the appropriate activity coefficient. The mean activity coefficient of an ion ($\gamma_{\pm}$) in solution can be measured, and there are now extensive tables for mono- and divalent ions (Hammer and Wu 1972; Staples and Nuttall 1977;

Goldberg and Nuttall 1978). While these values are all for 25 °C, the mean activity coefficient at other temperatures can be calculated by using the Debye-Hückel equation (see Baumgarten 1981). This paper also gives equations to calculate the mean activity coefficient at 25 °C for the most common physiological salt solutions based on the equations derived by Pitzer and Mayorga.

While the mean activity coefficient ($\gamma_{\pm}$) can be measured, this is not so with the single ion activity coefficient (γ_{+}, γ_{-}). To estimate it some nonthermodynamic assumptions have to be made, and there is no single agreed method for doing so (see Ammann 1986; Blatter et al. 1992). The group at the ETH in Zürich (Ammann 1986) have used the definition obtained from the Deby-Hückel theory which sets,

$$\log \gamma_{+} = \left|\frac{z^{+}}{z^{-}}\right|^{*} \log \gamma_{\pm}$$

$$\log \gamma_{-} = \left|\frac{z^{-}}{z^{+}}\right|^{*} \log \gamma_{\pm}$$

$\gamma_{\pm}$ is actually the mean activity coefficient for a molal solution. Concentrations in physiology are defined in terms of molarity, and it would be strictly accurate to use the mean molarity activity coefficient ($y_{\pm}$) for these solutions. However, it is the molal activity coefficients that are tabulated, and in the normal range of physiological concentrations the two are practically identical (see Blatter et al. 1992).

While the Nicolsky-Eisenman equation gave a good fit to the measured data in any one solution, it could not be used to predict the effects of changes of the ionic concentration as the selectivity coefficients were concentration dependent (Blatter and McGuigan 1988). This limits the practical application of the equation in the field of ion selective microelectrodes.

The relationship between the measured potential U and the concentration of the ion is logarithmic if the electrode is ideal. If, however, the electrode also reacts to other ions then the relationship between U and the logarithm of the concentration is no longer linear. Because of this nonlinearity, the concentration scale is truncated and is no longer logarithmic (see Fig. 7 in Chapman et al. 1983).

The electrodes are calibrated in a solution of known concentration and ionic strength and they measure the Mg^{2+} activity in this solution. It has often been debated whether the results should be expressed in activity or concentration, but provided that the ionic strength of the calibrating solutions are similar, it does not really matter for the mean activity coefficients for Mg^{2+} will also be similar. If there are large variations in ionic strength, this will not be so, and the results should be expressed in activity. In whatever form the results are expressed, concentration or activity, the composition of the calibrating solutions should be stated and when appropriate the activity coefficient used. It should also be made clear whether this is the mean or single ion activity coefficient (see also Blatter et al. 1992).

Construction of Ion Selective Microelectrodes

It is axiomatic that every laboratory has its own method for the manufacture of these electrodes, usually arrived at by trial and error. It is not intended to give a detailed account here but only to point out the features we think are important to obtain functioning electrodes. Details are given in Chapman et al. (1983) and in Blatter and McGuigan (1986). We use borosilicate glass (GC120F-10, Clark Electromedical Instruments, Reading, UK) pulled on a vertical Narishige puller to give microelectrodes which if filled with 3 *M* KCl would have a resistance of around 20 MΩ. We no longer wash or clean the glass because we have found it makes no difference to the performance of the electrodes. We routinely bevel our electrodes as it increases the likelihood that the electrodes will work, presumably because it increases the tip area (see plate 1 in Hess et al. 1982). Silanization is by the oven method of Tsien and Rink (1980) in which a few drops of *N*-trimethysilyldimethylamine (TMSDMA) were added to a dish containing the electrodes heated to a temperature of 200 °C. Initially chlorosilane was used, but this is highly corrosive and also not such an effective silanizing agent (Munoz et al. 1983). After silanization the electrodes can be stored for several weeks in a desiccator.

The tips are back-filled with resin using a Hamilton syringe, and care is taken to keep the column of resin as short as possible to prevent fluctuations induced by changes in bath temperature and solution level (Vaughan-Jones and Kaila 1986). After filling, the electrodes are stored over night, tip down, and during this time the resin diffuses into the tip. Small bubbles in the resin do not seem to interfere with the electrode performance and are no longer removed. Back-filling is with a syringe and again a bubble between the resin and the column of fluid in the shank is of no significance. Following the method of J. A. Coles a chlorided sliver wire is then sealed into the shank with shellac.

Electrodes filled with ETH 1117 are stored with their tips immersed in 100 m*M* $MgCl_2$ for a few days, which improves their performance. With the new resin ETH 5214 the electrodes have to be used immediately, as they deteriorate with time.

Performance of the Microelectrodes

ETH 1117

This resin is not highly specific as it also reacts to Ca^{2+}, Na^+, and K^+, and what this meant for the measurement of Mg^{2+} is shown in Fig. 3. Extracellularly, where the Ca^{2+} concentration is around 2 m*M* the response of the electrode remains flat because the electrode is reacting to the Ca^{2+} ions. Thus, it is not possible to use these electrodes to measure extracellular free Mg^{2+}.

In an intracellular situation in which the calcium concentration is less than 20 μ*M* (the contamination in our "calcium free" solution), the interference is

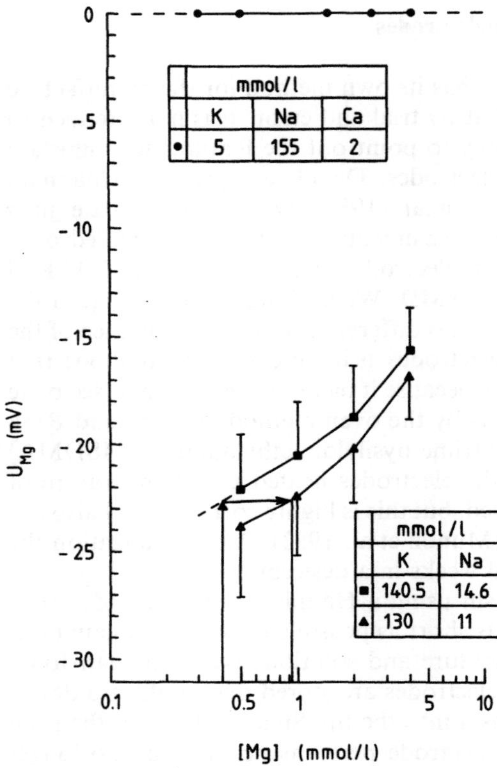

Fig. 3. Response of Mg^{2+} sensitive microelectrodes filled with ETH 1117 to extracellular and intracellular solutions. The response to the extracellular solution is for one electrode, and the Ca^{2+} concentration was constant throughout at 2 m*M*. Response to the intracellular solutions is the mean ± SD for seven electrodes. (Modified from Fig. 4 in Blatter and McGuigan 1988, with permission from Karger, Basel)

from Na^+ and K^+. Typical mean calibration curves in the presence of the measured mean intracellular concentrations of $[Na]_i$ and $[K]_i$ in ferret heart (Chapman et al. 1983; Reverdin et al. 1986) as well as the lowest measured values of $[Na^+]_i$ and $[K^+]_i$ (Blatter and McGuigan 1988) show that the deviation from linearity is significant. This reduces the response of the electrode between 5 and 0.5 m*M* Mg^{2+} from the expected 29.5 mV to 7.5 mV.

A more important disadvantage is that should $[Na^+]_i$ and/or $[K^+]_i$ change, so would the calibration curve. This is shown by the shift in the calibration curves to the right as the concentations of Na^+ and K^+ are reduced. As can be seen, the same magnesium signal U_{Mg} would give two different magnesium concentrations depending on the calibration curve chosen This property limits the application of the resin, as any change in the intracellular ionic concentrations (occurring when investigating the regulation of intracellular Mg^{2+}) would also change the calibration curve.

ETH 5214

Figure 4 compares the response of the new and the old resin in our standard background solution. The improvement is obvious, and the response between 4 and 0.5 m*M* Mg^{2+} is almost equal to the theoretical response (dashed line). However, ETH 5214 still reacts to Ca^{2+}, and in the presence of millimolar concentrations of Ca^{2+} the response to Mg^{2+} is flat, i.e. like its predecessor it cannot be used to measure extracellular free Mg^{2+}.

Response to Changes in Na^+ and K^+

Unlike ETH 1117 large changes in either Na^+ or K^+ do not influence the calibration curve of the new resin. In Fig. 5A reducing K^+ from 140.5 to 100 m*M* and increasing Na^+ from 14.6 to 40 m*M* as could occur during the calcium paradox in heart muscle (Chapman and Tunstall 1987) did not influence the calibration curve. Reducing $[Na^+]_o$ to zero causes a corresponding reduction in $[Na^+]_i$ in heart muscle (Chapman et al. 1983). As shown in

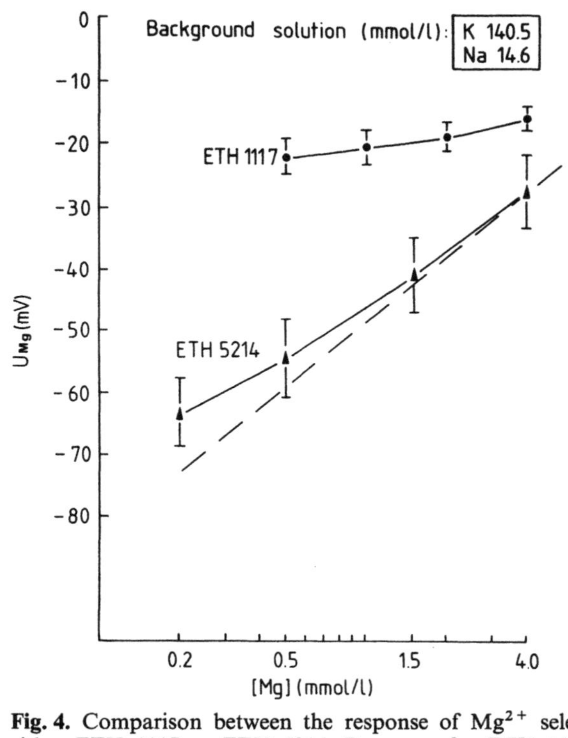

Fig. 4. Comparison between the response of Mg^{2+} selective microelectrodes filled with either ETH 1117 or ETH 5214. Response for ETH 1117 is the mean ± SD from seven electrodes, that for ETH 5214 for eight electrodes. *Dashed line* is the theoretical response for an electrode that only reacts to Mg^{2+}. Temperature 25 °C

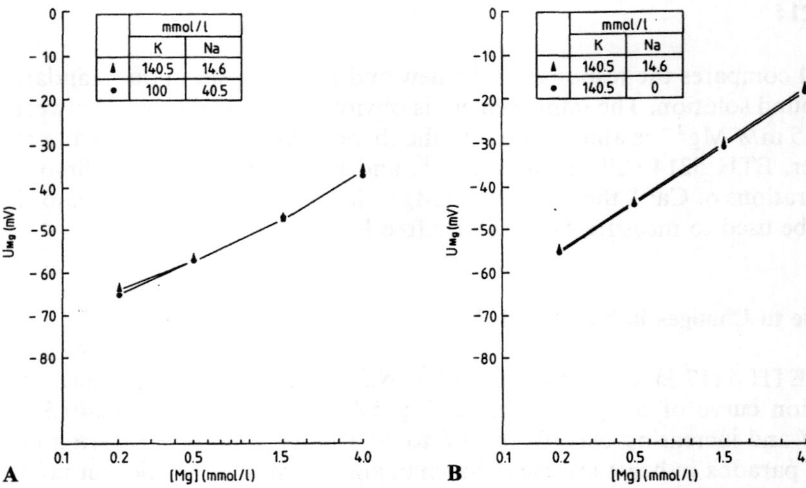

Fig. 5 A, B. Response of ETH 5214 filled microelectrodes to changes in K^+ and Na^+. **A** K^+ decreased from 140.5 to 100 m*M* and Na^+ simultaneously increased from 14.6 to 40 m*M*. **B** K^+ maintained constant but Na^+ decreased to zero. Temperature was 25 °C. Similar results were obtained in three other electrodes

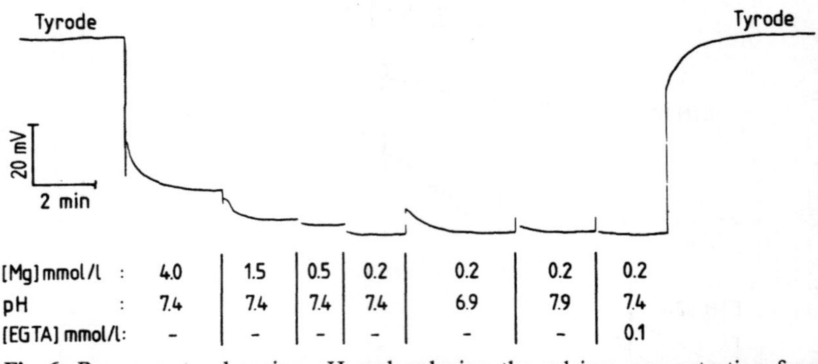

Fig. 6. Response to changing pH and reducing the calcium concentration from around 20 m*M* to about 10 n*M*. Temperature 25 °C. Similar results were obtained in three other electrodes

Fig. 5 B reducing the Na^+ concentration to zero while maintaining the K^+ concentration constant, also caused no change in the calibration curves.

Alterations of the external milieu can cause changes in pH_i and $[Ca]_i$ (Chapman et al. 1983; Blatter and McGuigan 1991). Figure 6 shows that changing the pH from 6.9 to 7.9 did not change the response of the electrode. The contamination level of calcium in our calibrating solutions is around 20 μ*M* higher than would be found intracellularly. However, the addition of 0.1 m*M* EGTA, which reduces the Ca^{2+} concentration to less than 10 n*M*, had no effect on the response. These tests show that ETH 5214 can not only measure resting levels of $[Mg^{2+}]_i$ much more accurately than its predecessor but it may also, because of the lack of response to Na^+ and K^+, be used to investigate the factors regulating $[Mg^{2+}]_i$.

Resistance of the Electrodes

The resistance of the electrodes is in the order of 20–30 GΩ, so that impedance matching is necessary. The buffer amplifier Analog Devices 311 J is ideally suited for experiments on ventricular muscle. Although bulky, it is almost impossible to damage, but unfortunately, it is no longer freely available.

Response Time of the Electrodes

The total response time of the electrodes is made up of the time to change the solution and the actual time response of the electrodes. In our rapid perfusion system the mean time response for 0% to 95% of the final signal on changing from 1.5 to 0.5 m*M* Mg^{2+} was 118 s ± 31 s (4 electrodes). It is thus clear that the electrodes can only be used to study mechanisms that are occurring over a time interval of minutes. If the tips are made larger, the response time of the electrodes is reduced, but they can no longer be inserted into the cell. Because of the slow time response of the ion selective microelectrode, the potential changes during an action potential are not accurately recorded. This means that subtraction of this signal from the signal of the 3 *M* KCl electrode would not result in an accurate estimation of the ionic concentration. To get round this problem both signals are filtered using low pass filters, and the filtered signals then subtracted to give the ionic concentration (Thomas 1978; Lee and Dagostino 1982; Vassalle and Lee 1984).

Measurement Problems with Ion Selective Microelectrodes

While useful information can be obtained with ion selective microelectrodes, the technique has its drawbacks. There are problems of obtaining properly functioning Mg^{2+} selective electrodes, and even functioning electrodes can give problems with calibration and drift. To reduce the time to obtain functioning electrodes they are initially screened. Since both the ion selective microelectrode and the 3 *M* KCl electrode have to be adequately inserted into the cell to obtain an accurate estimation of the $[Mg^{2+}]_i$, some test to check impalement is desirable. Each of these points will now be discussed.

Screening of Electrodes

After the electrodes are manufactured (ETH 5214) or after storage in 100 m*M* $MgCl_2$ (ETH 1117), they are examined to see what the calibration looks like.

Despite our attempts at trying to standardise our manufacturing procedure, from time to time a large proportion of the electrodes just do not work. During this initial screening it is also possible to discard electrodes with excessive drift or those that are extremely noisy.

Calibration

Calibration is carried out in a separate chamber whose temperature is thermostatically controlled (Blatter and McGuigan 1991). Although calibration and measurement are carried out in two different chambers with possibly different off-set potentials, this is not an actual problem since in each chamber Tyrode's solution is defined as zero. What is actually measured is the difference between this and either the calibrating solution or the cytosol (see appendix in Blatter and McGuigan 1988). We calibrated using a 3 *M* KCl electrode to minimize changes in junction potentials when changing from one calibrating solution to an another. Unless the chloride concentration is held constant, there will be changes in the potential of a Ag/AgCl earth electrode which can markedly alter the measured potential. In the calibration sufficient time has to be allowed for the signal to reach a steady state, which for Mg^{2+} electrodes can sometimes take up to 20 min or longer.

Since the response of the electrodes is temperature dependent, calibration and the experiment should be carried out at the same temperature. Calibration is routinely carried out before and after an impalement. This is necessary, for during the impalement the characteristics of the electrodes can change. Tissue attached to the tip of the electrode after an impalement can also give an off-set potential. If this is removed by washing the tip in running water, the calibration sometimes returns to its original form. The acceptable magnitude of the change in the calibration curves before and after an impalement is a matter of opinion. Blatter and McGuigan (1986) accepted a maximum difference of 1 mV. Another way to express the variation is to accept a difference in concentration as obtained from the calibration curves before and after the impalement. For the new magnesium resin ETH 5214 a difference of 0.4 m*M* was accepted.

Since most ion selective microelectrodes do react to other ions, it has become customary to calibrate the electrodes in a solution mimicking the intracellular concentration. It is essential that these concentrations are the measured concentrations of the interfering ions, or large errors can occur, especially in the case of the magnesium sensor ETH 1117 (see later).

Drift

This can be minimized by a suitable choice of electrode. If it does occur then again a maximal accepted drift has to be agreed upon. We accept a drift of 1 mV/min (Chapman et al. 1983).

Adequate Impalement

This is an important topic for in order to arrive at the magnesium signal U_{Mg}, the membrane potential (E_m) has to be subtracted from the potential of the Mg^{2+} electrode (V_{Mg}). If either one or both are not adequately impaled in the cell the estimated Mg^{2+} concentration can be spuriously high or low.

The ion selective microelectrodes may sometimes slightly damage the cell membrane, for the potential of the electrode drifts down for periods of up to 30 min (see Blatter and McGuigan 1986). This sealing-in of the electrode means that the Mg^{2+} signal increases over this time and reduces the estimated $[Mg^{2+}]_i$. This was especially important with ETH 1117 for with that resin a few mV made a large difference to the estimated Mg^{2+} concentration (see Fig. 3). Another problem is that a steady signal from the ion selective microelectrode is no guarantee that the electrode is measuring the correct concentration of Mg^{2+}. In other words it is easy to obtain high values. One way round this difficulty is to depolarize the muscle by increasing the K^+ concentration, in our case from 5 to 14 m*M*. If both electrodes are in the cell they should both be depolarized by the same amount, and the U_{Mg} signal should not change. However, implicit in this method is the assumption that depolarization per se does not change the Mg^{2+} concentration.

It is normal to accept a maximal deviation of 1 mV from such a test. A deviation of 1 mV with a depolarization of 20 mV means that 95% of the true signal is being measured. If it is assumed that it is only the ion selective microelectrode that is not registering the whole potential, then it is possible to estimate what this deviation would mean in terms of concentration. For ETH 1117 if only 95% of the signal was measured the mean concentration would increase from 0.4 to 2.5 m*M*; for ETH 5214 the mean concentration increase would be from 0.85 to 1.5 m*M*. The main reason for the large difference between the two is due to the greater slope of ETH 5214 (see Fig. 4).

These calculations show that even with a deviation on the U_{Mg} signal of 1 mV the range in concentrations is large for the sensor ETH 1117, and it might be asked if it is worthwhile carrying out such a test. However, it must be pointed out that this is the maximum deviation, and in many experiments there is no change in potential during the test (see Fig. 2 in Blatter and McGuigan 1986), and since there is no deviation it can be accepted that both electrodes are adequately impaled. As such measurements are accumulated, the mean resting value becomes known, and during an experiment it is possible to check the impalement against the calibration curve. If the value seems either too high or too low the electrodes can be re-inserted.

Presentation of Data Obtained with Ion Selective Microelectrodes

It has been common in the presentation of measurements obtained with ion selective microelectrodes to convert the measured potential signal into a concentration (Fig. 1 B) and to present the concentrations as a normally distributed set (see Fig. 5 for Na^+ measurements in Chapman et al. 1983). However, the measured potential for an ideal electrode is logarithmically related to the concentration (Fig. 2), and the question arises: what is the normally distributed set, concentration or logarithm of the concentration? In an attempt to answer this question we collaborated with Dr. C. H. Fry and his group at St. Thomas's Hospital in London, for both our laboratories have accumulated numerous measurements with ion selective microelectrodes in cardiac muscle. The Kolmogorov-Smirnov test (Sokal and Rohlf 1969) was used to check for a normal distribution, and it was possible to show that it was the logarithm of the concentration that was the normally distributed set, the distribution of the concentration being skew. This has important consequences for the calculation of the mean intracellular ionic concentrations, for calculation on the assumption that the concentrations are normally distributed gives values that are too high. Recalculation of Fry's Mg^{2+} measurements (1986) in heart muscle, for instance, reduced the mean value from 1.91 to 1.69 m*M* (Fry et al. 1990). A corollary to this is that frequency plots of the concentration will show a tail. It is also interesting to note that pH has always been assumed to be normally distributed, and the mean value for pH has always been calculated from the pH values (Waddell and Bates 1969). This led to a dichotomy in the literature in that pH or $-\log a_H$ was assumed to be normally distributed, but it was the concentrations of the other ions that were assumed to be normally distributed. Since it has now been shown that it is the logarithm of the concentration that is normally distributed, this dichotomy no longer exists.

Measurement of $[Mg^{2+}]_i$ in Ventricular Muscle

Reparation

We have routinely used small bundles of ferret trabeculae (diameter around 250 μm) mounted in a rapid perfusion chamber (see Chapman et al. 1983; Reverdin et al. 1986). The temperature was maintained at 25 °C, a temperature chosen originally because it simplified the calibration of the electrodes. A typical measurement using the magnesium sensor ETH 1117 from Blatter and McGuigan (1986) is shown in Fig. 7 in which from above downwards are shown the membrane potential (E_m), the Mg^{2+} potential (U_{Mg}), and the potential of the Mg^{2+} selective microelectrode (V_{Mg}). In such measurements it is *absolutely* essential to make sure that a steady state has been reached, which can take up to 30 min as is shown by the slow downward drift of U_{Mg} in Fig. 7. U_{Mg} is the difference between V_{Mg} and E_m, and in order to test whether both electrodes were in the cell, the cells were depolarized by increasing $[K^+]_o$ from 5 to 14 m*M*. If both electrodes were correctly impaled, then both should see the same change in E_m and U_{Mg} should not change, as was indeed the case in this impalement. This test was routinely carried out for each and every impale-

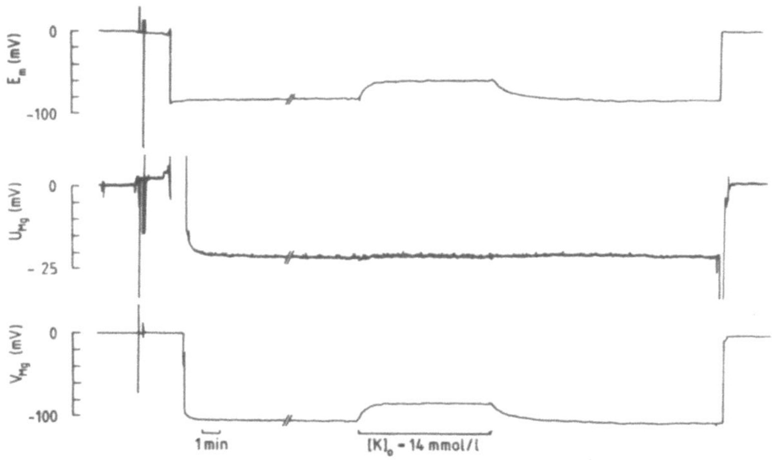

Fig. 7. Typical intracellular measurement for Mg^{2+} using ETH 1117. From above downward are shown membrane potential (E_m), Mg^{2+} potential (U_{Mg}), and Mg^{2+} selective microelectrode potential (V_{Mg}). The double lines mark a 22 min recording gap. (From Fig. 2 of Blatter and McGuigan 1986, with permission)

ment, and in the majority of cases the deviation was minimal. In these experiments the measured concentrations of Na^+ and K^+ were used for the calibration solutions, and sufficient time was always allowed for the electrodes to seal in.

Thirteen measurements in seven experiments could be successfully carried out, and the mean value for $[Mg^{2+}]_i$ was 0.4 m*M*, this mean value being estimated from the mean pMg_i value (see Fig. 3 B in Blatter and McGuigan 1986).

This value is highly dependent on the values of K^+ and Na^+ chosen for the calibrating solution (cf. Fig. 3). If the $[K^+]_i$ and $[Na^+]_i$ values in the cells in which $[Mg^{2+}]_i$ was actually measured were less than the mean values, then the estimated $[Mg^{2+}]_i$ would increase. To set an upper limit to the measured $[Mg^{2+}]_i$, Mg^{2+} selective microelectrodes were calibrated in solutions containing the mean measured values for $[Na^+]_i$ and $[K^+]_i$ in ferret ventricle and solutions containing the lowest measured values of $[Na^+]_i$ and $[K^+]_i$ (see Fig. 3). If all the measurements had been carried out in cells containing the lowest measured $[Na^+]_i$ and $[K^+]_i$, the estimated $[Mg^{2+}]_i$ would increase to 0.9 m*M*. Thus, $[Mg^{2+}]_i$ in heart muscle lies, in round figures, between 0.5 and 1 m*M* (Bladder and McGuigan 1988).

The new Mg^{2+} sensor ETH 5214 does not suffer from the disadvantage of reacting to K^+ and Na^+, and its calibration curve does not change when Na^+ and K^+ concentrations are widely varied. Measurements of $[Mg^{2+}]_i$ in ferret ventricular muscle and frog skeletal muscle using the criteria described in Blatter and McGuigan (1986) for successful impalements gave mean $\pm$ SD for pMg_i of 3.07 ± 0.28 (24 experiments, 32 impalements) and 3.03 ± 0.42 (38 experiments), respectively. These correspond to mean values of 0.85 and 0.93 m*M* for heart and skeletal muscle (Blatter et al. 1989). Thus, our original Mg^{2+} measurements in ferret ventricular muscle have been confirmed by the new sensor.

Comparison with Other $[Mg^{2+}]_i$ Measurements in the Literature

Garfinkel et al. (1986) and Blatter and McGuigan (1988) have recently summarized $[Mg^{2+}]_i$ in vertebrate tissue. Tables 1, 2, and 3 give these values plus twelve additional values arranged in chronological order for skeletal, heart, and smooth muscle, and the results are plotted as histograms in Fig. 8. In Table 1 the value of Günther and Dorn (1971) has been converted to concentration using the activity coefficient given by the authors. In Table 2 the corrected value for Fry (1986) has been given (see Fry et al. 1990).

All the results for smooth muscle are under 1 m*M*, and the recent results from heart and skeletal muscle lie in the vicinity of 1 m*M*. As has previously been pointed out (Blatter and McGuigan 1988) the distribution for heart and skeletal muscle is bimodal, twelve of the results being over 2 m*M*. Six of these

Table 1. $[Mg^{2+}]_i$ in skeletal muscle

Reference	$[Mg^{2+}]_i$ (m*M*)	Method	Species
Gilbert 1960	1.1	uptake of Mg^{28}	frog
Nanninga 1961	3.4	calculation	frog
Günther and Dorn 1971	1.6–1.9	macroelectrode cell homogenate	rat
Endo 1977	0.8	Ca^{2+} release from SR[a]	frog
Cohen and Burt 1977	3–4.4	^{31}P-NMR	frog
Moore 1978	1.0	^{31}P-NMR	frog
Dawson et al. 1978	>2.5	^{31}P-NMR	frog
Gupta and Moore 1980	0.6	^{31}P-NMR	frog
Hess et al. 1982	3.3	microelectrodes	frog
Miledi et al. 1982	<1.0	arsenazo III	frog
Maughan 1983	0.2	diffusible Mg^{2+}	frog
Close and Lännergren 1984	3.5	arsenazo III	frog
Lopez et al. 1984	3.8	microelectrodes	frog
Maughan and Recchia 1985	0.6	diffusible Mg^{2+}	frog
Alamo et al. 1986	3.7	microelectrodes	rabbit
Alvarez-Leefmans et al. 1986	0.8	microelectrodes ($[Mg^{2+}]_o=0$)	frog
	1.7	microelectrodes ($[Mg^{2+}]_o=0.5$m*M*)	frog
Kushmerick et al. 1986	1.5	^{31}P-NMR	cat
Baylor et al. 1986	1–2	arsenazo III, antipyrylazo III	frog
MacDermott 1987	0.5	microelectrodes	rat
Lopez et al. 1990	1.6	microelectrodes	swine
Blatter et al. 1989	0.9	microelectrodes	frog

[a] SR, Sarcoplasmic reticulum

Table 2. $[Mg^{2+}]_i$ in heart muscle

Reference	$[Mg^{2+}]_i$ (m*M*)	Method	Species
England et al. 1967	<1	Mg^{2+} dependence of citrate/isocitrate	rat
Steenbergen et al. 1978	0.6	Mg^{2+} dependence of glycolysis	rat
Paradise et al. 1978	3.6	Mg^{2+} efflux in zero Ca^{2+}	rabbit
Wu et al. 1981	2.5	^{31}P-NMR	guinea pig
Hess et al. 1982	3.0	microelectrodes	ferret
	3.1		sheep
	3.5		Purkinje fibres sheep
Gupta et al. 1984	0.8	^{31}P-NMR	guinea pig
Kushmerick et al. 1986	1.4	^{31}P-NMR	rat
Blatter and McGuigan 1986	0.4	microelectrodes	ferret
Fry 1986	1.7	microelectrodes	ferret
Gupta and Wittenberg 1987	0.5	^{31}P-NMR	myocytes
Kirschenlohr et al. 1988	1.5	^{19}F-NMR	ferret
Borchgrevink et al. 1989	0.5	^{31}P-NMR	rat
Headrick and Willis 1989	0.8	^{31}P-NMR	rat
Kirkels et al. 1989	0.4	^{31}P-NMR	rat
Murphy et al. 1989 a	0.5	Furapta	chick cells
Murphy et al. 1989 b	0.9	^{19}F-NMR	rat
Blatter et al. 1989	0.9	microelectrodes	ferret

Table 3. $[Mg^{2+}]_i$ in smooth muscle

Reference	$[Mg^{2+}]_i$ (m*M*)	Method	Species
Palaty 1971	0.1	Mg^{2+} efflux	rat tail artery
Dillon 1986	0.4	^{31}P-NMR	porcine carotid artery
Kushmerick et al. 1986	0.5	^{31}P-NMR	rabbit bladder
	0.4		rabbit uterus
Nakayama et al. 1987	0.2	^{31}P-NMR	guinea pig intestine

values were obtained with microelectrodes (Hess et al. 1982; Lopez et al. 1984; Alamo et al. 1986), and these high values can be accounted for by the fact that the authors did not measure the $[K^+]_i$ and $[Na^+]_i$ in their preparations and did not allow sufficient time for the electrodes to seal in.

Three values were obtained with NMR studies. The studies of Cohen and Burt (1977) and Dawson et al. (1978) were carried out at 4 °C under anaerobic conditions, and under these conditions the Mg^{2+} concentration could well be higher. Recalculation of the results of Wu et al. (1981) by Garfinkel and Garfinkel (1984) reduced the value from 2.5 to 1.6 m*M*. Of the remaining three values, the value of Paradise et al. (1978) was obtained after Ca^{2+} free perfusion, a condition similar to calcium paradox in heart (Grinwald and Nayler 1981). Nanninga (1961) calculated the free Mg^{2+} concentration but ignored

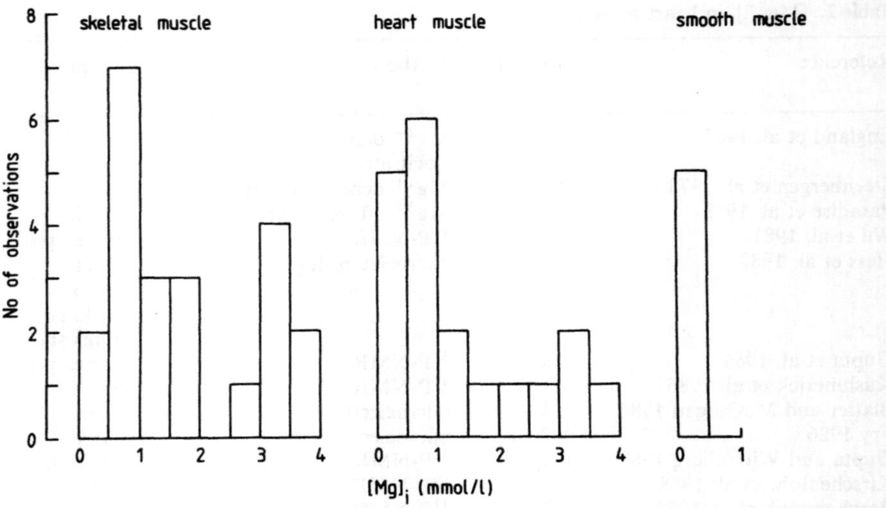

Fig. 8. Histogram of the values for $[Mg^{2+}]_i$ in muscle tissue tabulated in Tables 1, 2 and 3

binding to mitochondria and the cell membrane (see Garfinkel et al. 1986). The final value over 2 m*M* is that measured by Close and Lännergren (1984) using arsenazo III, but as has recently been pointed out intracellular calibration of this dye is difficult (Baylor et al. 1986). In summary, if these twelve values are excluded for the reasons mentioned above, the distribution of values for $[Mg^{2+}]_i$ in muscle tissue is around 1 m*M*.

Conclusions

In the past there has been much uncertainty about the level of free Mg^{2+} in vertebrate muscle cells for different methods apparently gave different values. However, recent careful measurements by various methods show that Mg^{2+} is around 1 m*M*. This variation in the literature around 1 m*M* could very well be physiological.

The new Mg^{2+} sensor ETH 5214 is ideally suited for the study of the factors regulating intracellular Mg^{2+}, but because of its sensitivity to Ca^{2+}, it cannot be used to measure Mg^{2+} extracellularly. It is to be hoped that Simon's group at the ETH in Zürich will soon be able to overcome this disadvantage. The new sensor together with the recent NMR (Kirschenlohr et al. 1988; Levy et al. 1988) and fluorescent methods (Raju et al. 1989) usher in a new era in Mg^{2+} research, and it is to be expected that major advances will now occur in this to some extent neglected field.

Acknowledgements. Excellent technical help was provided during the course of these experiments by Marlis Jordi and Daniel Lüthi. Darani Hapangama during her vacation helped to screen the electrodes. Prof. Clamann and Rosemary McGuigan made helpful comments on the manuscript. Work supported by the Swiss National Science Foundation grants no 3.320-0.82 and 3.210-0.85. Support was also provided by the Swiss Society for Magnesium Research. A. Buri had a stipendium from the Stanley Thomas Johnson Foundation.

References

Alamo L, Lopez JR, Papp L, Sreter FA (1986) Simultaneous measurements by means of ion-specific electrodes of free intracellular ionized calcium and magnesium in rabbit skeletal muscle. Muscle Nerve 9:472–474

Altura BM (1988) Ischemic heart disease. Magnesium 7:57–67

Alvarez-Leefmans FJ, Gamino SM, Giraldez F, Gonzalez-Serratos H (1986) Intracellular free magnesium in frog muscle fibres measured with ion-selective micro-electrodes. J Physiol (Lond) 378:461–483

Ammann D (1986) Ion-selective micro-electrodes. Principles, design and application. Springer, Berlin Heidelberg New York

Baumgarten CM (1981) A program for calculation of activity coefficients at selected concentrations and temperatures. Comput Biol Med 11:186–196

Baylor SM, Hollingworth S, Hui CS, Quinta-Ferreira ME (1986) Properties of the metallochromic dyes arsenazo III, antipyrylazo III and azo 1 in frog skeletal muscle fibres at rest. J Physiol (Lond) 377:89–141

Blatter LA, McGuigan JAS (1986) Free intracellular magnesium concentration in ferret ventricular muscle measured with ion selective micro-electrodes. Q J Exp Physiol 71:467–473

Blatter LA, McGuigan JAS (1988) Estimation of the upper limit of the free magnesium concentration measured with Mg-sensitive microelectrodes in ferret ventricular muscle: (1) use of the Nicolsky-Eisenman equation and (2) in calibrating solutions of the appropriate concentration. Magnesium 7:154–165

Blatter LA, McGuigan JAS (1991) Intracellular pH regulation in ferret ventricular muscle: the influence of metabolic substrates and the role of Na/H exchange. Circ Res (in press)

Blatter LA, Buri A, McGuigan JAS (1989) Free intracellular magnesium concentration in isolated ferret ventricular muscle and in frog skeletal muscle measured with ion-selective microelectrodes containing the new magnesium sensor ETH 5214. J Physiol 418:154P

Blatter LA, Fry CH, McGuigan JAS (1992) Ion selective microelectrodes: relevant physiological chemical properties of solutions and the principles of application of the electrodes to intracellular measurement. Prog Biophys Mol Biol (to be submitted)

Borchgrevink PC, Bergen AS, Bakoy OE, Jynge P (1989) Magnesium and reperfusion of the ischemic rat heart as assessed by ^{31}P-NMR. Am J Physiol 256:H195–H204

Brady H, Ryan M, Horgan J (1987) Magnesium: the forgotten cation. Ir Med J 80:250–253

Chapman RA, Suleman IH (1986) Changes in a_{Na} and pH_i in isolated ferret ventricular muscle on repletion of the bathing Ca following a period of Ca-deprivation. J Physiol (Lond) 381:120P

Chapman RA, Tunstall J (1987) The calcium paradox of the heart. Prog Biophys Mol Biol 50:67–96

Chapman RA, Coray A, McGuigan JAS (1983) Sodium/calcium exchange in mammalian ventricular muscle: a study with sodium-sensitive microelectrodes. J Physiol (Lond) 343:253–276

Close RI, Lännergren JI (1984) Arsenazo III calcium transients and latency relaxation in frog skeletal muscle fibres at different sarcomere lengths. J Physiol (Lond) 355:323–344

Cohen SM, Burt CT (1977) ^{31}P nuclear magnetic relaxation studies of phospocreatine in intact muscle: determination of intracellular free magnesium. Proc Natl Acad Sci USA 74:4271–4275

Corkey BE, Duszynski J, Rich TL, Matschinsky B, Williamson JR (1986) Regulation of free and bound magnesium in rat hepatocytes and isolated mitochondria. J Biol Chem 261:2567–2574

Dawson MJ, Gadian DG, Wilkie DR (1978) Muscular fatigue investigated by phosphorus nuclear magnetic resonance. Nature 274:4271–4275

Dillon PF (1986) Free intracellular Mg^{2+} in relaxed and contracted porcine carotid arteries. J Physiol (Lond) 371:161P
Elin RJ (1988) Magnesium metabolism in health and disease. Dis Mon 4:163–219
Endo M (1977) Calcium release from the sarcoplasmic reticulum. Physiol Rev 57:71–108
England PJ, Denton RM, Randle PJ (1967) The influence of magnesium ions and other bivalent metal ions on the aconitase equilibrium and its bearing on magnesium ions by citrate in rat heart. Biochem J 105:32C–33C
Fry CH (1986) The actions of extracellular magnesium on isolated mammalian ventricular myocardium. J Physiol (Lond) 372:32P
Fry CH, Hall SK, Blatter LA, McGuigan JAS (1990) Analysis and presentation of intracellular measurements obtained with ion-selective microelectrodes. Exp Physiol 75:187–198
Garfinkel L, Garfinkel D (1984) Calculation of free Mg^{2+} concentration in adenosine 5′-triphosphate containing solutions in vitro and in vivo. Biochemistry 23:3547–3552
Garfinkel L, Altschuld RA, Garfinkel D (1986) Magnesium in cardiac energy metabolism. J Mol cell Cardiol 18:1003–1013
Gilbert DL (1960) Magnesium equilibrium in muscle. J Gen Physiol 43:1103–1118
Goldberg RN, Nuttall RL (1978) Evaluated activity and osmotic coefficients for aqueous solutions: the alkaline earth metal halides. J Phys Chem Ref Data 7:263–310
Grinwald PM, Nayler WG (1981) Calcium entry in the calcium paradox. J Mol Cell Cardiol 13:867–880
Günther T, Dorn F (1971) Die intrazelluläre Mg-Ionenaktivität in verschiedenen Säugetierzellen. Z Naturforsch 26B:176–177
Gupta RK, Moore RD (1980) ^{31}P NMR studies of intracellular free Mg^{2+} in intact frog skeletal muscle. J Biol Chem 255:3987–3993
Gupta RK, Wittenberg BA (1987) NMR measurements of cytosolic free magnesium in cardiac myocytes. J Mol Cell Cardiol 19 [Suppl IV]:S51
Gupta RK, Gupta P, Moore RD (1984) NMR studies of intracellular metal ions in intact cells and tissues. Annu Rev Biophys Bioeng 13:221–246
Hammer WJ, Wu YC (1972) Osmotic coefficients and mean activity coefficients of univalent electrolytes in water at 25 °C. J Phys Chem Ref Data 4:1047–1099
Headrick J, Willis RJ (1989) Cytosolic $[Mg^{++}]$ in simulated hypoxic, or ischaemic rat heart. J Mol Cell Cardiol 21 [Supp II]:S40
Hess P, Metzger P, Weingard R (1982) Free magnesium in sheep, ferret, and frog striated muscle at rest measured with ion selective micro-electrodes. J Physiol (Lond) 333:173–188
Hu Z, Bührer T, Müller M, Rusterholz B, Rouilly M, Simon W (1989) Intracellular magnesium ion selective microelectrodes based on a neutral carrier. Anal Chem 61:574–576
Kirkels JH, van Echteld CJA, Ruigrok TJC, Meijler FL (1989) Intracellular magnesium during ischemia and reperfusion. J Mol Cell Cardiol 21 [Suppl II]:S110
Kirschenlohr HL, Metcalfe JC, Morris PG, Rodrigo GC, Smith GA (1988) Ca^{2+} transient, Mg^{2+}, and pH measurements in the cardiac cycle by ^{19}F-NMR. Proc Natl Acad Sci USA 85:9017–9021
Kushmerick MJ, Dillon PF, Meyer RA, Brown TR, Krisanda JM, Sweeny HL (1986) 31P-NMR spectroscopy, chemical analysis, and free Mg^{2+} of rabbit bladder and uterine smooth muscle. J Biol Chem 261:14420–14429
Lanter F, Erne D, Ammann D, Simon W (1980) Neutral carrier based ion-selective electrode for intracellular magnesium activity studies. Anal Chem 52:2400–2402
Lauter DP (ed) (1989) Magnesium deficiency – pathogenesis, prevalence, and strategies for repletion. Am J Cardiol 18 [Suppl G]
Lee CO, Dagostino M (1982) Effects of strophanthidin on intracellular Na activity and twitch tension of constantly driven canine Purkinje fibres. Biophys J 40:185–198
Levy LA, Murphy E, Raju B, London RE (1988) Measurement of cytosolic free magnesium ion concentration by 19F NMR. Biochemistry 27:4041–4048
Lopez JR, Alamo C, Caputo, Vergara J, DiPolo (1984) Direct measurement of intracellular free magnesium in frog skeletal muscle using magnesium-sensitive microelectrodes. Biochim Biophys Acta 804:1–7

Lopez JR, Sanchez V, Ryan JF, Sreter FA, Allen PD (1990) The effects of magnesium on myoplasmic $[Ca^{2+}]$ in malignant hyperthermia susceptable swine. Anesthesiology (in press)
MacDermott M (1987) The intracellular concentration of free magnesium in rat skeletal muscle. J Physiol (Lond) 394:14P
Maughan D (1983) Diffusible magnesium in frog skeletal muscle. Biophys J 43:75–80
Maughan D, Recchia C (1985) Diffusible sodium, potassium, magnesium, calcium, and phosphorus in frog skeletal muscle. J Physiol (Lond) 368:545–563
Miledi R, Parker I, Zhu PH (1982) Calcium transients evoked by action potentials in frog twitch muscle fibres. J Physiol (Lond) 333:655–679
Moore RD (1978) 31P NMR study of intracellular free Mg^{2+} in intact frog muscle. Fed Proc 37:1608
Munoz JL, Deyhimi F, Coles JA (1983) Silanization of glass in the making of ion-sensitive microelectrodes. J Neurosci Methods 8:231–247
Murphy E, Freudenrich CC, Levy LA, London RE, Lieberman M (1989a) Monitoring cytosolic free magnesium in cultured chicken heart cells by the fluorescent indicator Furaptra. Proc Natl Acad Sci USA 86:2981–2984
Murphy E, Steenbergen C, Levy LA, Raju B, London RE (1989b) Cytosolic free magnesium levels in ischemic rat heart. J Biol Chem 264:5622–5627
Nakayama S, Chihara D, Uno Y, Tomita T (1987) Intracellular Mg^{2+} in the taenia of guinea-pig caecum. J Muscle Res Cell Motil 8:283
Nanninga LB (1961) Calculation of free magnesium, calcium and potassium in muscle. Biochim Biophys Acta 54:338–344
Page E, Polimeni PI (1972) Magnesium exchange in rat ventricle. J Physiol (Lond) 224:121–139
Paradise NF, Beeler GW, Vissher MB (1978) Magnesium net fluxes and distribution in rabbit myocardium in irreversible contracture. Am J Physiol 234:C115–C121
Palaty V (1971) Distribution of magnesium in the arterial wall. J Physiol (Lond) 218:353–368
Raju B, Murphy E, Levy LA, Hall RD, London RE (1989) A fluorescent indicator for measuring cytosolic free magnesium. Am J Physiol 256:C540–C548
Reverdin EC, Illanes A, McGuigan JAS (1986) Internal potassium activity in ferret ventricular muscle. Q J Exp Physiol 71:451–465
Review (1987) Potassium magnesium depletion: is your patient at risk of sudden death. Am J Med 82 [Suppl 3A]
Shils ME (1988) Magnesium in health and disease. Annu Rev Nutr 8:429–460
Sokal RT, Rohlf FJ (1969) Biometry. The principles and practice of statistics in biological research. Freedman, San Francisco
Staples BR, Nuttall RL (1977) The activity and osmotic coefficients of aqueous calcium chloride at 298.15 K. J Phys Chem Ref Data 6:385–407
Steenbergen C, Deleeuv G, Williamson JR (1978) Analysis of control of glycolysis in ischemic hearts having heterogeneous zones of anoxia. J Mol Cell Cardiol 10:617–639
Thomas RC (1978) Ion-sensitive intracellular microelectrodes. How to make and use them. Academic Press, London
Tsien RY, Rink RJ (1980) Neutral carrier ion-selective microelectrodes for measurement of intracellular free calcium. Biochim Biophys Acta 599:623–638
Vassalle M, Lee CO (1984) The relationship among intracellular sodium activity, calcium and strophanthidin inotrophy in canine cardiac Purkinje fibres. J Gen Physiol 83:287–307
Vaughan-Jones RD, Kaila K (1986) The sensitivity of liquid sensor, ion-selective microelectrode to changes in temperature and solution level. Pflugers Arch 406:641–644
Waddell WJ, Bates RG (1969) Intracellular pH. Physiol Rev 49:285–329
White RE, Hartzell HC (1989) Magnesium ions in cardiac function. Regulator of ion channels and second messengers. Biochem Pharmacol 38:859–867
Wu ST, Pieper GM, Salhany JM, Eliot RS (1981) Measurement of free magnesium in perfused and ischemic arrested heart muscle. A quantitative phosphorus-31 nuclear magnetic resonance and multiequilibria analysis. Biochemistry 20:7399–7403

Voltage-Gated Ionic Channels: Diversity and Modulation by Mg^{2+} [1]

Arthur M. Brown [2] and Atsuko Yatani

Thirty years ago, the masterful voltage clamp studies of Hodgkin and Huxley (1952) and Cole (1949) focussed attention on one voltage-gated Na^+ current and one voltage-gated K^+ current. Today we know of at least five cDNAs encoding Na^+ channels (Numa 1988; Goldin et al. 1986; Trimmer et al. 1989), at least four cDNAs encoding K^+ channels, of which the *Shaker* cDNA can be alternately spliced in five different ways (Jan and Jan 1989), and at least two cDNAs encoding the α_1 subunit of the Ca^{2+} channel (Tanabe et al. 1988; Mikami et al. 1989; Perez-Reyes et al. 1989). It is likely that many more genes will be found because of the many different known phenotypes, especially inwardly rectifying, ligand-gated (G_K, Na^+, ATP) K^+ channels, and low or high threshold Ca^{2+} channels. Diversity has clearly displaced uniformity, yet despite this the secondary structures and membrane topologies predicted from the various known primary sequences share common features. In the most popular model of Na^+ and Ca^{2+} channels (Numa 1988), homologous domains numbering 1 to 4 are composed of six transmembrane α-helical segments one of which, S4, has an R-X-X-R repeat which may be the voltage sensor. The K^+ channel cDNAs, which were discovered later, encode only one domain but are more amenable to recombinant methods and are likely to be the most widely studied as a result. Historically, the breakthrough came with the cloning of the *Electrophorus* Na^+ channel (Noda et al. 1984) with which the first part of this review begins. In the second part, since this symposium has Mg^{2+} effects on ionic channels as its theme, modulatory phenomena related to the role of intracellular Mg^{2+} on muscarinic atrial K^+ channels will be described.

Diversity of Voltage-Gated Ionic Channels

The Na^+ channel is a large, transmembrane glycoprotein and in some tissues is made up of several subunits (Table 1), although currents can be expressed by transcripts from the large α subunit alone. The voltage-gated Na^+ channel

[1] This work was supported in part by National Institutes of Health grants NS 23877, HL36930, HL37044, and HL39262 to A. M. Brown
[2] Department of Molecular Physiology and Biophysics, 1 Baylor Plaza, Baylor College of Medicine, Houston, Texas 77030, USA

cDNA was first isolated from the electric eel using immunoscreening of a cDNA expression library and comparing the deduced amino acid (a.a.) sequence from partial clones with the sequence from tryptic peptides (Noda et al. 1984). The deduced a.a. sequence did not show a signal peptide, and the amino terminus was on the cytoplasmic side of the membrane. The predicted secondary structure and membrane topology were derived from hydropathy plots, and from these a model (Fig. 1) was proposed. The model has four homologous domains, each domain containing five hydrophobic segments (S1, S2, S3, S5, S6) and one positively charged segment (S4), with an arginine or lysine at every third position which has been proposed as the voltage sensor. The best functional evidence for this idea comes from a recent study in which the arginines of putative domain one were mutated to neutral or negatively

Table 1. Comparison of different Na^+ channel proteins

Species	Subunit mass and composition
Electrophorus electricus	260–300 K_d
Rat brain	260 K_d (α), 36 K_d ($\beta1$), 33 K_d ($\beta2$)
Rat skeletal muscle	260 K_d (α), 39 K_d (β)
Chick cardiac muscle	230–270 K_d (α)

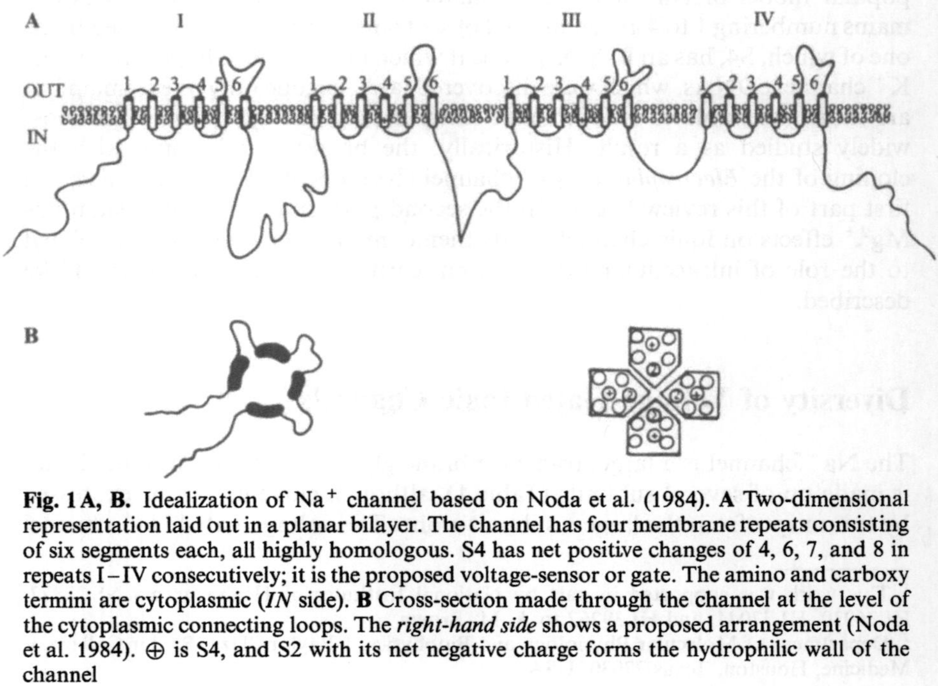

Fig. 1 A, B. Idealization of Na^+ channel based on Noda et al. (1984). **A** Two-dimensional representation laid out in a planar bilayer. The channel has four membrane repeats consisting of six segments each, all highly homologous. S4 has net positive changes of 4, 6, 7, and 8 in repeats I–IV consecutively; it is the proposed voltage-sensor or gate. The amino and carboxy termini are cytoplasmic (*IN* side). **B** Cross-section made through the channel at the level of the cytoplasmic connecting loops. The *right-hand side* shows a proposed arrangement (Noda et al. 1984). ⊕ is S4, and S2 with its net negative charge forms the hydrophilic wall of the channel

Table 2. Similarity of predicted amino acid sequences among different Na^+ channels

		MW	Percentage similarity			
			I	II	III	Eel
Na^+ channel I	2009 (delI 1998)	228758	100	87	85	61
Na^+ channel II	2005	227840		100	87	62
Na^+ channel III	1951	221375			100	61
Na^+ channel Eel	1820					100

charged amino acids individually or in combinations. The mutants showed significant changes in conductance–voltage relationships of the type consistent with S4 being the activation gate (Stuhmer et al. 1989).

Transcripts from the *Electrophorus* cDNA did not express in *Xenopus* oocytes, and the next step forward involved Na^+ channel cDNAs from rat brain which expressed currents in oocytes. Three distinct cDNA clones were isolated from rat brain using eel cDNA probes and had overall structures that were similar to the eel channel (Table 2).

The four homologous domains are highly conserved (nearly identical) among the three Na^+ channels from rat brain, and as for eel muscle, in each domain there are five hydrophobic transmembrane segments. S1, S2, and S3 contain a few charged residues, S4 contains four to eight arginine or lysine residues at every third position, and S5 and S6 are very hydrophobic. There are 5 of 6 conserved N-glycosylation sites on the outside, and all cAMP-dependent phosphorylation sites (PKA) are on the inside.

In vitro RNA transcripts were synthesized from cDNAs from rat Na^+ channeles II and III cloned into transcription-competent vectors. The transcripts were injected into *Xenopus* oocytes, and 3–6 days later Na^+ channels were detected using a two-microelectrode, voltage-clamp technique. What was observed is that from a holding potential of −100 mV and test potentials from −50 to +60 mV, inward currents were produced that were tetrodotoxin (TTX) sensitive. The currents were reduced by replacement of external Na^+ with tetraethylammonium (TEA), tetramethyl ammonium (TMA), or sucrose. Maximal Na^+ (I_{Na}) current occurred at test potentials between −10 and 0 mV, and 50% of the channels were inactivated by prepulses to −50 mV. Activation was fast, but inactivation was slower than expected. Both processes, but especially inactivation, were quickened by co-injection of low molecular weight RNA, raising the possibility of β subunit effects or some difference in processing (Krafte et al. 1988).

We have recently studied single Na^+ III channels expressed in *Xenopus* oocytes and found that the gating could switch even in the absence of cytoplasmic modulators. This demonstrates that the protein itself can have long periods in which it resides in different conformations that manifest different kinetics (Moorman et al. 1989).

Another phenomenon of interest is the fact that subconductance states are very common for these expressed Na^+ channels. This now appears to be true of native Na^+ channels as well. The previous idea of two-state channels, either non-conducting or conducting, is the exception rather than the rule for all types of voltage-gated channels.

Ca^{2+} Channels

Ca^{2+} channels are functionally more diverse and structurally more complex than Na^+ channels. The skeletal muscle T-tubule Ca^{2+} channel has been studied most extensively and is a pentameric structure consisting of α_1, α_2, β, γ, and δ components (Table 3) (Perez-Reyes et al. 1989).

Dihydropyridines (DHPs) have been used as Ca^{2+} blockers and were used to purify the DHP receptor (DHPR) from skeletal muscle. The purified DHPR has been incorporated into phospholipid bilayers, and active, DHP-sensitive Ca^{2+} channels were recovered that retained physiological and pharmacological properties of a 20 pS functional L-type channel. For example, the protein was phosphorylated by PKA, which increased voltage-dependent opening probability (Flockerzi et al. 1986).

The α_1 subunit has 1873 a.a. (212018 daltons), and its hydrophobicity profile is similar to that of Na^+ channels, giving a predicted general "organization" reminiscent of Na^+ channels. The cDNA for the α_2 subunit encodes for 1106 a.a. (125018 daltons). There is a 26-a.a. signal sequence present at the aminoterminus which is extracellularly located. There are 18 potential N-glycosylation sites (Asn-X-Ser/Thr) and 2 cAMP-dependent phosphorylation sites. The α_2 subunit has no homology to known ionic channel or receptor proteins.

The α_1 and α_2 subunits are expressed differently in various tissues. Northern blot analyses with RNA from various tissues revealed the α_1 transcript to be present predominantly in skeletal muscle although weak signals came from the aorta and heart. It was not deteced in the ileum or brain. The α_2 transcript was, however, detected in the RNA of all the tissues examined.

Recently, we have stably transformed mouse L cells with α_1 DHPR subunit (Perez-Reyes et al. 1989). These cells have no evidence of α_2 and/or δ subunits. They expressed DHP binding with the same K_d as found in rabbit skeletal muscle, although the density of receptors is far lower. Calcium currents were also expressed, and these had extremely slow kinetics (Perez-Reyes et al. 1989).

Table 3. Subunit structure of the skeletal muscle T-tubule dihydropyridine receptor (DHPR)

α_1	175 K_d	DHP binding subunit	Phosphorylated by PKA
α_2	143 K_d	Glycosylated	
β	54 K_d		Phosphorylated by PKA
γ	30 K_d	Glycosylated	
δ	27 K_d	Glycosylated	

PKA, cAMP-dependent protein kinease

Numa's group has recently cloned the α_1 DHPR from heart and expressed convincing Ca^{2+} currents in *Xenopus* oocytes. These currents were increased when α_2 transcripts were co-injected (Mikami et al. 1989).

K^+-Channels

Since K^+ channels have not yet been biochemically purified, very little independent structural information is currently available. Recently, a family of A-type channels from *Drosophila* has been isolated by cDNA cloning, and their function has been assessed by K^+ channel expression in *Xenopus* oocytes (Jan and Jan 1989).

Shaker mutants of *Drosophila* shake their legs when anesthetized with ether. Voltage-clamp studies in pupal and larval muscle showed that a fast, transient K^+ current (I_A) was affected (Salkoff and Wyman 1981). There are several *Shaker* alleles known that correlate with different A current amplitudes, kinetics, or voltage-dependence. All this was indirect evidence that the *Shaker* locus encodes a structural component of the A channel.

Several *Shaker* mutations have been caused by chromosomal translocation. Such translocations have been used to localize the *Shaker* mutations cytologically to a region on the X chromosome (polytene band 16F). A cDNA clone was available that hybridized to 16F and was used to initiate a bidirectional "walk" of 350 kilobases to cover the region of the *Shaker* locus.

DNA fragments covering the "walk" from +30 to +120 kb were used as probes to screen several cDNA libraries. Three classes of cDNA clones were isolated, and the isolates were shown to be encoded at the *Shaker* locus (by in situ hybridization to region 16F on polytene chromosomes). The different isolates (ShA1; ShA2; ShB1; ShC1) were derived by alternative splicing from a large (>65 kb) primary transcript (Jan and Jan 1989).

The cDNA sequence of ShA1 was 2928 base pairs with an open reading frame of 1848 nucleotides encoding 616 a.a. The molecular weight 70 200 compared with 65–75 K_d from dendrotoxin (an A channel blocker) binding studies. From hydropathy analysis, the hydrophilic domains at the amino and carboxyl termini joined with a central core of six hydrophobic stretches of about 19 a.a. These were modelled as α-helical transmembrane segments, and one was an S4-like, arginine-rich segment with seven basic a.a. As for voltage-gated Na^+ and Ca^{2+} channels, the topology of the channel showed no signal peptide, the amino terminus was cytoplasmic, and ShA1 encoded a monomer (616 a.a.) analogous to one domain of the Na^+ or Ca^{2+} channels. A functional A channel may be formed as a homomultimer, but the number of subunits has not been determined.

Several different proteins were encoded by the *Shaker* locus. A family of transcripts was generated by alternative RNA splicing of a large precursor RNA encoded by the *Shaker* gene. This may be the mechanism for the expression of A channel subtypes that differed in their voltage dependence, kinetics, cellular and subcellular distribution, and temporal expression. ShA and ShB cDNA clones were transcribed in vitro, and the resulting mRNA was injected in *Xenopus* oocytes. Transient outward currents (characteristic of A-type channels) could be detected. A single *Shaker* mRNA sufficed to direct the synthesis and surface expression of functional A channels (Jan and Jan 1989).

Recently, we have used expression cloning to isolate a unique K^+ channel from rat brain (Frech et al. 1989). This channel is more similar to SHAB, a putative K^+ channel, than it is to *Shaker* (Salkoff and Wyman 1981). Interestingly, it expresses the slow rise time, steady current, and voltage dependence of a delayed rectifier K^+ channel.

Modulation of Voltage-Gated Ionic Channels

The best understood modulatory pathways involve the seven helix membrane spanning receptors that include noradrenergic receptors (Dixon et al. 1986), signal transducing guanine nucleotide binding or G proteins, and Ca^{2+} channels, although there is a rash of data accumulating on Na^+ and K^+ channels as well. Two pathways are involved, and the G proteins are a branch point for them: indirect via cytoplasmic second messengers and direct or membrane-delimited via the G proteins themselves. The best-studied channel in the latter category is the ACh-activated or muscarinic atrial K^+ (K^+[ACh]) channel (Brown and Birnbaumer 1988, 1990). We took advantage of the fact that once the K^+[ACh] channel has been activated either by guanosine 5′-(γ-thio)-triphosphate (GTPγS) activation of endogenous G_K or by addition of preactivated pertussis toxin (PTX) substrate from human erythrocytes, activated by GTPγS (G_K^*) or the α-subunit of G_K, activated by GTPγS (α_i^*-3), then Mg^{2+} can be removed from the internal solution and the channel remains activated for many hours. Therefore, we could evaluate effects of changing concentrations of Mg^{2+} at relatively short times (<10 min) where the effects are on the K^+[ACh] channel and not on the G protein.

Intracellular Mg^{2+} is known to block outward currents in voltage-gated and ligand-gated, inwardly rectifying K^+ channels (Horie and Irisawa 1987). Fig. 2 Aa shows that, as previously reported, intracellular Mg^{2+} is required for K^+[ACh] channel activation by GTPγS (Fig. 2 Ab); however, once activation has occurred, openings persist after withdrawal of Mg^{2+} (Kurachi et al. 1986). The $[Mg^{2+}]_i$ requirements for G protein activation of K^+[ACh] channels are shown in Fig. 2 B. K_d was 3 μ*M*. Both preactivated α_K^* (α_i-3) and recombinant α_i^*-3 were effective in the absence of intracellular Mg^{2+} (Fig. 2 b, c), and native a_i^*-2 was as effective as native α_i^*-3. Moreover, the channel open time and conductance at potentials more negative than the equilibrium potential for K^+ (E_K) in the presence and absence of Mg^{2+} are the same for all these conditions of activation (Table 4).

As previously reported (Horie and Irisawa 1987), when K^+[ACh] channels were activated by GTPγS after removal of intracellular Mg^{2+} the single channel conductance showed an ohmic current–voltage (I-V) relationship between −90 and +70 mV (Fig. 3). This was also the case after activation with native

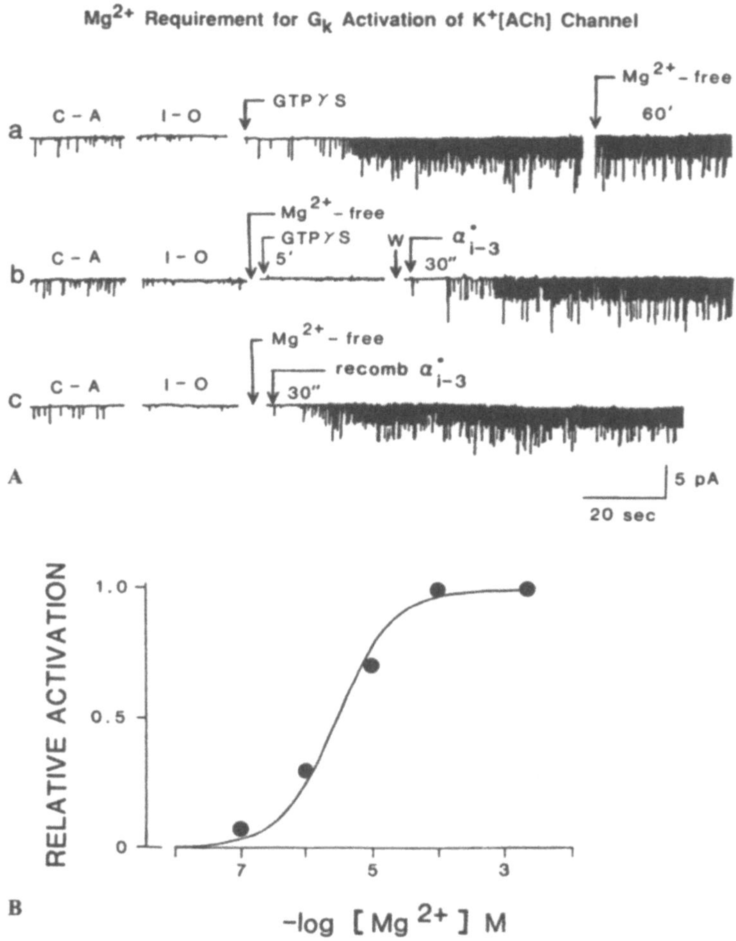

Fig. 2A, B. Effects of internal Mg^{2+} on the atrial K^+ channels. **A** Atrial K^+ channel currents activated with GTPγS (*a*), native α_i^*-3 (*b*), and recombinant α_i^*-3 (*c*) in inside-out membrane patches excised from adult guinea pigs. Records were first obtained in the cell-attached configuration (*C-A*) then inside-out (*I-O*) configuration with symmetrical isotonic K^+ solutions at a holding potential of −80 mV. The membrane patches show little basal activation in the C-A configuration without agonist in the pipette. Times on top of each recording are minutes (′) or seconds (″) elapsed between the addition of the test agent to the bathing solution and the illustrated recordings. *a* Effects of intracellular Mg^{2+} on GTPγS (100 µ*M*) activated currents. *b* Without intracellular Mg^{2+}, GTPγS did not activate the channel, but native α_i^*-3 activated the channel. *c* Recombinant α_i^*-3 also activates the channel without intracellular Mg^{2+}. **B** Concentration-dependent effects of intracellular Mg^{2+} on muscarinic activation of single channel K^+ currents in inside-out membrane patches. Data were obtained at a fixed carbachol concentration of 10 µ*M* in the patch pipette and at 100 µ*M* GTP in the bath solution. *NP*, where *P* is the proportion of open time over 30 s and *N* is the number of channels was measured, at each concentration of Mg^{2+} 1 min after application and was normalized to the maximum value obtained with 2 m*M* Mg^{2+}. Concentration-response curve was fit by a Michaelis-Menten method with a K_d of 3 µ*M*. Data were mean values from 3–5 experiments

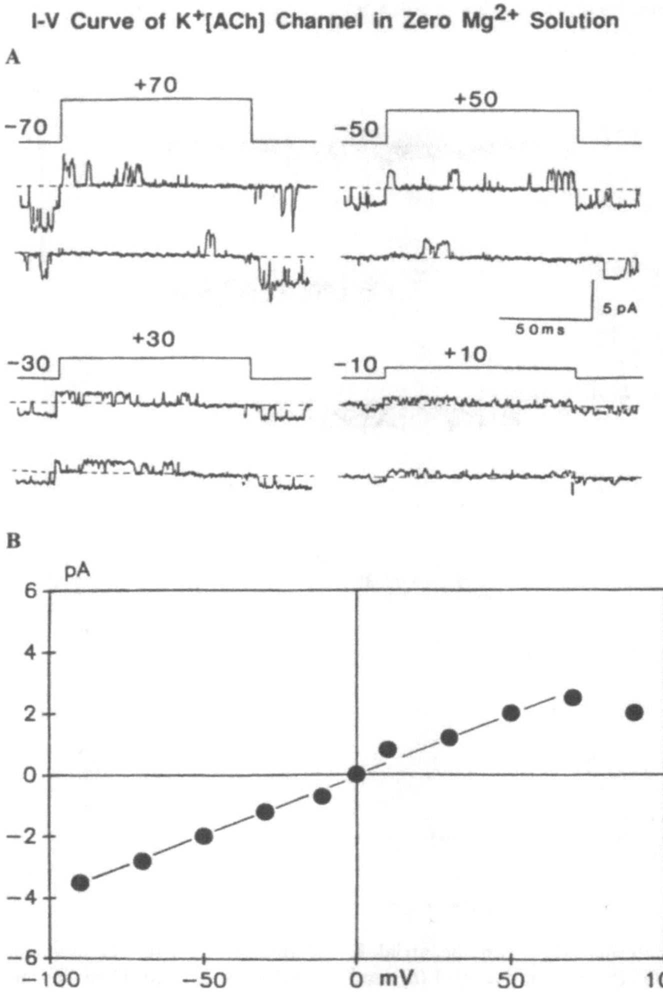

Fig. 3. A Single channel currents recorded from recombinant α_i^*-3 (500 p p*M*) activated atrial K^+ channels in inside-out membrane patches at indicated potentials in the absence of intracellular Mg^{2+}. Pulse protocols are shown above the current traces. Depolarizing test pulses are applied every 3 s. *Dottled line* indicates zero current levels. **B** Single channel current–voltage (I-V) relationships obtained in the absence of intracellular Mg^{2+}. Current amplitudes were measured as currents were activated by recombinant α_i^*-3 under steady-state conditions. *Straight line* was drawn by eye with single channel conductance of 40 pS between membrane potentials of −90 and +70 mV with a similar conductance. Currents activated with GTPγS, native α_i^*-2 or α_i^*-3, and recombinant α_i^*-3

Table 4. Effects of Mg^{2+}-free internal solution on single muscarinic atrial K^+ channel current. Mean open times of single atrial K^+ currents were activated by GTPγS, G_k^*, α_i^*-2 or α_i^*-3, or recombinant (r) α_i^*-3. Data were obtained from inside-out membrane patches in Mg^{2+}-free symmetrical K^+ solution at a membrane potential of -70 mV. In each case slope conductance between -90 and $+60$ mV was ~40 pS

Conditions	Mean open time (ms)	n
GTPγS	1.5 ± 0.2	5
G_k^*	1.6 ± 0.4	3
α_i^*-2 or α_i^*-3	1.7 ± 0.3	6
rα_i^*-3	1.6 ± 0.2	4

Values are mean ±SD

α_i^*-3 and recombinant α_i^*-3. Prior to removal of intracellular Mg^{2+}, inward rectification was prominent. Figure 3A illustrates outward and inward currents at different holding potentials in the absence of intracellular Mg^{2+}. The outward currents, which were rare in the presence of intracellular Mg^{2+}, appeared more frequently at the beginning of the depolarizing pulses and became less frequent during the pulse.

Using symmetrical 140 m*M* K^+ solutions, the conductance was 40 ± 2 pS (mean ± SD) at room temperature. Half-amplitude openings were frequent at positive potentials in Mg^{2+}-free solution, but their frequency relative to control in the presence of intracellular Mg^{2+} has not been examined. Horie and Irisawa (1987) have noted that the probability of opening of K^+[ACh] channels is strongly voltage dependent when channels were activated with either native α_i^*-3 or recombinant α_i^*-3. After producing steady activation 5–6 min after application, we measured the proportion of open time P during 100-ms pulses for N channels in the patch pipette, together called NP. Integrated NP values for 2-min periods were plotted as cumulative NP. The ratio of averaged NP at different potentials to maximum NP (at -80 mV) was plotted against test potentials. NP was markedly reduced at depolarized potentials, and the effects on averaged NP were similar when channels were activated with GTPγS, native α_i^*-3, or recombinant α_i^*-3.

Similar relaxation of whole-cell, ACh-activated, outward K^+ currents have been studied using microelectrode and whole-cell clamp measurements (Noma and Trautwein 1978; Endoh et al. 1985; Iijima et al. 1985).

A recent report (Szabo et al. 1988) claimed that preactivated native α_i^*-3 applied at nanomolar concentrations intracellularly did not produce whole-cell current relaxations consistent with those observed for muscarinic currents. This indicates that other currents must be activated because at the single channel level, endogenous G_K or α_i^*-3 and α_i^*-2 or exogenous recombinant α_i^*-3 all have the same effects in the presence or absence of intracellular Mg^{2+}.

Summary

Intracellular Mg^{2+} has multiple effects on K^+[ACh] channels. In biochemical studies, it binds to the $G_i\alpha$ subunit, the physiological activator, with high affinity (nanomolar K_d) and to one of the $\alpha\beta\gamma$ subunits with low (micromolar K_d) affinity (Gilman 1987). The high affinity sites are involved in the GTPase effect while the low affinity sites are involved in GDP dissociation, GTP binding, and dissociation of the α and $\beta\gamma$ subunits. Intracellular Mg^{2+} also blocks outward currents through the open channels at submillimolar levels. It will be of interest to compare its biochemical and physiological effects on G_K-gating of the K^+[ACh] channel.

Acknowledgements. We thank Judy Breedlove for her secretarial assistance.

References

Brown AM, Birnbaumer L (1988) Direct G protein gating of ion channels. Am J Physiol 23(3): H401–H410

Brown AM, Birnbaumer L (1990) Ionic channels and their regulation by G protein subunits. Annu Rev Physiol 52:197–213

Cole KS (1949) Dynamic electrical characteristics of the squid axon membrane. Arch Sci Physiol 3:253–258

Dixon RAF, Kobilka BK, Strader DJ, Benovic JL, Dohlman HL, Frielle T, Bolanowski MA, Bennett CD, Rands E, Diehl RE, Mumford RA, Slater EE, Sigal IS, Caron MG, Lefkowitz RJ, Strader CD (1986) Cloning of the gene and cDNA for mammalian beta-adrenergic receptor and homology with rhodopsin. Nature 321:75–79

Endoh M, Maruyama M, Iijima T (1985) Attenuation of muscarinic cholinergic inhibition by islet-activating protein in the heart. Am J Physiol 249: H309–H320

Flockerzi V, Oeken HJ, Hofman F, Pelzer D, Cavalie A, Trautwein W (1986) Purified dihydropyridine-binding site from skeletal muscle T-tubules is a functional calcium channel. Nature 323:66–68

Frech GC, VanDongen AMJ, Schuster G, Brown AM, Joho RH (1989) A novel rat brain potassium channel with delayed rectifier properties isolated through expression cloning. Nature 340:642–645

Gilman AG (1987) G proteins: transducers of receptor-generated signals. Annu Rev Biochem 56:615–649

Goldin AL, Snutch T, Lubbert H, Dowsett A, Marshall J, Auld V, Downey W, Fritz LC, Lester HA, Dunn R (1986) Messenger RNA coding for only the alpha subunit of the rat brain Na channel is sufficient for expression of functional channels in *Xenopus* oocytes. Proc Natl Acad Sci USA 83(19):7503–7507

Hodgkin AL, Huxley AF (1952) A quantitative description of membrane current and its application to conduction and excitation in nerve. J Physiol (Lond) 117:500–544

Horie M, Irisawa H (1987) Rectification of muscarinic K^+ current by magnesium ion in guinea pig atrial cells. Am J Physiol 253: H210–H214

Iijima T, Irisawa H, Kameyama M (1985) Membrane currents and their modification by acetylcholine in isolated single atrial cells of guinea pig. J Physiol 359:485–501

Jan LY, Lan YN (1989) Voltage-sensitive ion channels. Cell 56(1):13–25

Krafte DS, Snutch TP, Leonard JP, Davidson N, Lester HA (1988) Evidence for the involvement of more than one mRNA species in controlling the inactivation process of rat and rabbit brain Na channels expressed in *Xenopus* oocytes. J Neurosci 8(8):2859–2868

Kurachi Y, Toshiaki N, Sugimoto T (1986) Role of intracellular Mg^{2+} in the activation of muscarinic K^+ channel in cardiac atrial cell membrane. Pflugers Arch 407:572–574
Mikami A, Imoto K, Tanabe T, Niidome T, Mori Y, Takeshima H, Narumiya S, Numa S (1989) Primary structure and functional expression of the cardiac dihydropyridine-sensitive calcium channel. Nature 340:230–233
Moorman JR, Kirsch GE, Lacerda AE, Brown AM (1989) Fast and slow gating of sodium channels encoded by a single mRNA. Neuron 4:243–252
Noda M, Shimizu S, Tanabe T, Takai T, Kayano T, Ikeda T, Takahashi H, Nakayama H, Kanaoka Y, Minamino N, Kangawa K, Matsuo H, Rafftery MA, Hirose T, Inayama S, Hayashida H, Miyata T, Numa S (1984) Primary structure of electrophorus electricus sodium channel deduced from cDNA sequence. Nature 312:121–127
Noma A, Trautwein W (1978) Relaxation of ACh-induced potassium current in rabbit sinoatrial node cell. Pflügers Arch 377:193–200
Numa S (1988) A molecular view of neurotransmitter receptors and ionic channels. Harvey Lect Rev 83:121–165
Perez-Reyes E, Kim HS, Lacerda AE, Horne W, Wei X, Rampe D, Campbell KD, Brown AM, Birnbaumer L (1989) Induction of calcium currents by the expression of the $\alpha 1$-subunit of the dihydropyridine receptor from skeletal muscle. Nature 340:233–236
Salkoff L, Wyman R (1981) Genetic modification of potassium channels in *Drosophila Shaker* mutants. Nature 293:228–230
Stuhmer W, Conti F, Suzuki H, Wang XD, Noda M, Yahagi N, Kubo H, Numa S (1989) Structural parts involved in activation and inactivation of the sodium channel. Nature 339:597–603
Szabo G, Pang IH, Sternweis PC (1988) Activation of a K^+ current by exogenous G proteins in atrial myocytes. Biophys J 53:424a
Tanabe T, Beam KG, Powell JA, Numa S (1988) Restoration of excitation-contraction coupling and slow calcium current in dysgenic muscle by dihydropyridine receptor complementary DNA. Nature 336:134–139
Trimmer JS, Cooperman SS, Tomiko SA, Zhou J, Crean SM, Boyle MB, Kallen RG, Sheng Z, Barchi RL, Sigworth FJ, Goodman RH, Agnew WS, Mandel G (1989) Primary structure and functional expression of a mammalian skeletal muscle sodium channel. Neuron 3:33–49

ATP-Dependent K^+ Channels of Excitable Membranes

Peter R. Stanfield [1]

The discovery in heart muscle of K^+ channels closed by the binding of ATP (Noma 1983) has its roots in an earlier observation that when cells are metabolically exhausted, the K^+ permeability of the cell membrane rises substantially. Such a rise in permeability was known to occur in heart muscle, where the action potential becomes shorter during exhaustion (Trautwein et al. 1954); in skeletal muscle, where resting K^+ conductance can increase some 200-fold in rigor, resulting in inexcitability (Fink and Lüttgau 1976); and in neurones (Godfraind et al. 1970). In most cases the increase in K^+ permeability was initially attributed to activation of Ca^{2+}-dependent K^+ permeability (see for example Krnjevic and Lisiewicz 1972), which had already been identified by Whittam (1968) in red cells and by Meech and Strumwasser (1970) in neurones. Opening of such channels may indeed provide part of the conductance increase since ethylene glycol-bis(β-aminoethylether)*N,N,N′,N′*-tetraacetic acid (EGTA) injection (which will reduce free intracellular Ca^{2+}) reduces the conductance change (Fink et al. 1983). It is now clear that ATP-dependent K^+ channels supply a major part of the permeability increase at least in heart and skeletal muscle and possibly in certain neurones as well (Fujiwara et al. 1987; Mourre et al. 1989). In this chapter, I shall discuss the nature and function of ATP-dependent K^+ channels and the roles of intracellular Mg^{2+} in modulating channel open state probability (P_{open}), either directly or through binding of nucleotides and in reducing K^+ efflux by channel blockage at positive membrane potentials.

Distribution of ATP-Dependent K^+ Channels

The discovery of ATP-dependent K^+ channels in the heart was quickly followed by their demonstration in pancreatic B cells (Cook and Hales 1984) and in skeletal muscle (Spruce et al. 1985). More recently, such channels have also been shown to be present in neurones (Ashford et al. 1988), especially in nerve terminals (Ben Ari and Krnjevic 1989), and in smooth muscle (Davies et al. 1989a; Standen et al. 1989). There seems little reason for not assuming that they are even more widespread in their distribution. Furthermore, the phenomenon of regulation of ion channels by ATP binding may itself be a more

[1] Department of Physiology, University of Leicester, PO Box 138, Leicester LE1 9HN, UK

general property, since Ca^{2+} channels of sarcoplasmic reticulum (Smith et al. 1985) and Ca^{2+}-dependent 'non-selective' cation channels of tissues as diverse as the pancreatic islets (Sturgess et al. 1987) and the ascending limb of the renal loop of Henle (Paulais and Toulon 1989) are also ATP dependent. Even voltage-dependent K^+ channels of the squid axon may have certain of their properties modulated by ATP binding (Clay and Szuts 1988).

Discovery of ATP-Dependent Channels

ATP-dependent K^+ channels were discovered using the method of patch clamp with excised inside-out membrane patches, where the cytoplasmic face of the membrane could be exposed to different concentrations of ATP. Single-channel records from frog skeletal muscle are shown in Fig. 1. These records were made in approximately physiological K^+ concentrations ($[K^+]_o$ 2.5 m*M*; $[K^+]_i$ 120 m*M*) but in the absence of either ATP or internal Mg^{2+}. Figure 1 also illustrates unitary current-voltage relations, which show outward rectification at physiological $[K^+]_o$ (where $[K^+]_o \ll [K^+]_i$) but even in the absence of internal Mg^{2+} some measure of inward rectification at high $[K^+]_o$. The unitary conductance of these channels appears to vary from tissue to tissue, although conductances are similar in heart, skeletal muscle and pancreatic B cells in

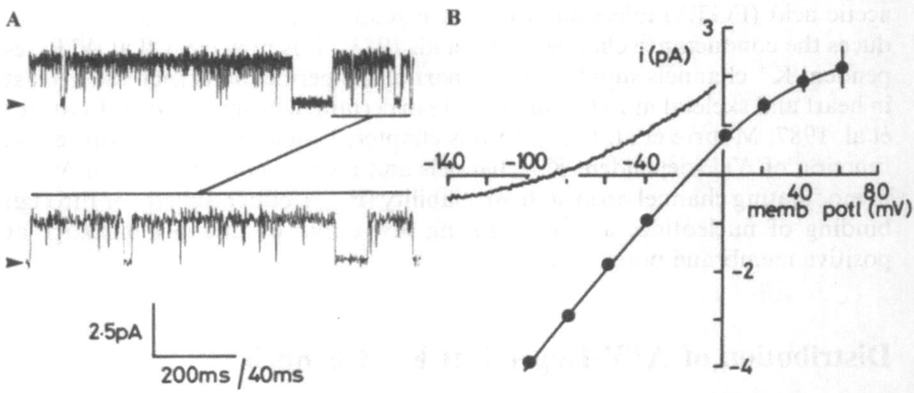

Fig. 1. A Records of unitary current through ATP-dependent K^+ channels from a patch of skeletal muscle membrane with $[K^+]_o = 2.5$ m*M*, $[K^+]_i = 120$ m*M*. Records replayed from digital storage on a video recorder through an 8-pole Bessel low-pass filter, with the −3 dB frequency set at 2 kHz. The lower record shows the part of the upper record indicated by the horizontal lines on the faster time-base. By courtesy of Dr NW Davies. **B** Current–voltage relations for ATP-dependent K^+ channel of skeletal muscle with $[K^+]_o$ 2.5 (*upper*) and 60 m*M* (*lower*). The relation in 2.5 m*M* K^+ was measured with voltage ramps in four patches, ensemble averaging parts of records where one channel was open, after subtraction of capacity currents associated with charging the recording pipette and of leakage currents through the seal. The relation in 60 m*M* K^+ is the mean from five patches, and the *vertical bars* indicate ± standard error of the mean. From Spruce et al. (1987), *Journal of Physiology*, by permission

which the conductance is around 50 pS in high external $[K^+]$ and around 15 pS in physiological $[K^+]_o$ at room temperature (see, for example, Spruce et al. 1987). Unitary conductance appears higher in neurones (Ashford et al. 1988) and in smooth muscle (Davies et al. 1989a; unitary conductance 135 pS in 60 m*M* $[K^+]_o$ at 22 °C). The kinetic properties of channels are broadly similar among different tissues, with more than one open state and at least three closed states (see, for example, Spruce et al. 1985, 1987; Stanfield 1987), although the affinity for ATP appears to vary. There is considerable variation in the reported stoichiometry of ATP binding (see below, Fig. 3).

Function of ATP-Dependent K^+ channels

The opening of ATP-dependent K^+ channels under conditions of exhaustion is strongly suggestive of a link between their open state probability and the metabolic state of the cell in whose membrane they reside. Certainly in pancreatic B cells and insulin-secreting cell lines, this linkage is quite clear (Ashcroft et al. 1984; Rorsman and Trube 1985) and essential in allowing glucose uptake and metabolism, resulting in channel closure by an increase in intracellular [ATP], to stimulate the secretion of insulin. A similar linkage between glucose metabolism and an increased firing of action potentials has been suggested for glucose-responsive neurones (Ono et al. 1982) of the hypothalamus (Stanfield 1987; Ashford et al. 1989). Such a role, linking excitability or K^+ permeability to metabolic state, has also been envisaged in heart and skeletal muscle, though an understanding of the role has been complicated by the high affinity of the channels for ATP and by the remembrance that cellular ATP may be buffered by creatine phosphate and creatine kinase, changing little at least in the cell as a whole over a period of muscle contraction (Carlson and Siger 1960). Nonetheless, roles in control of action potential duration, of excitability and of the release of K^+ from muscle fibres to act as a vasodilator have been envisaged (see, for example, Spruce et al. 1987). The high density with which these channels are found argues against their being opened only in extremes of exhaustion.

ATP-Depenent K^+ Channels Are also Opened by Peptides

More recently, it has been shown that the channels can be opened by the action of peptides or other factors, so that it seems possible that, in some tissues at least, the binding of ATP, instead of being the primary controlling factor, is rather a requirement for holding the channels in a closed state from which they may be opened by the action of other agents. The peptide galanin is known to open ATP-dependent K^+ channels in pancreatic B cells (and to inhibit insulin release in consequence, De Weille et al. 1988). In the nervous system, one of the functions of galanin is to modulate transmitter release from mossy fibre termi-

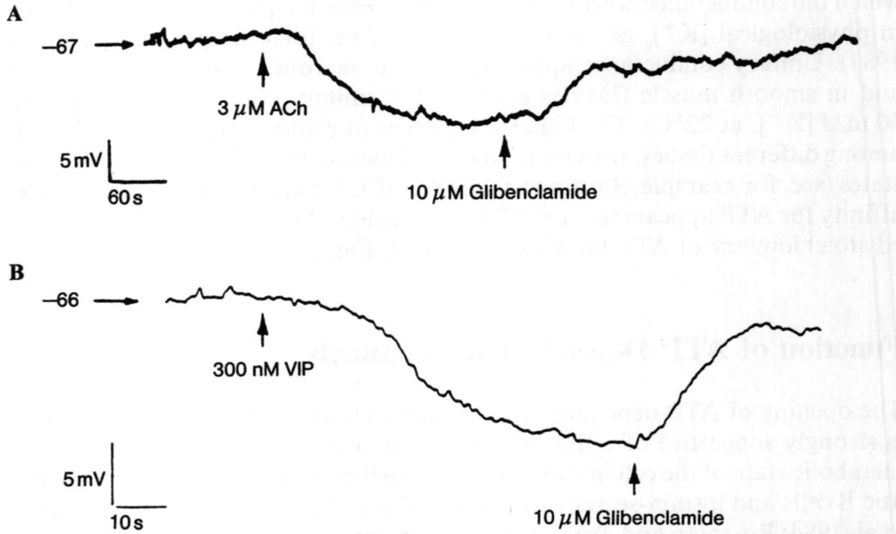

Fig. 2 A, B. Records of resting potential of smooth muscle cells recorded with impaling microelectrodes from mesenteric artery. **A** The hyperpolarising action of acetylcholine (endothelium-dependent) is reversed by application of glibenclamide, a specific blocker of ATP-dependent K^+ channels. **B** The hyperpolarising action of vasoactive intestinal peptide is also reversed by glibenclamide treatment. Reproduced from Standen et al. (1989) Science 245:117–180, with permission; copyright 1989 by the AAAS

nals of the hippocampus by opening ATP-dependent K^+ channels there (Ben Ari and Krnjevic 1989). Vasoactive intestinal peptide (VIP) seems to exert a major part of its vasodilator action by opening these channels in smooth muscle of the mesenteric artery (Fig. 2 B; Standen et al. 1989). The recently described endothelium-derived hyperpolarising factor (EDHP; Taylor and Weston 1988) may well act on the same class of K^+ channels – at least the endothelium-dependent hyperpolarising action of acetylcholine (known to release EDHP through activation of M_1 muscarinic receptors; Taylor and Weston 1988) is reversed by the ATP-dependent K^+ channel blocker glibenclamide (Standen et al. 1989; Fig. 2 A).

ATP-Dependent K^+ Channels Bind ATP

ATP-dependent K^+ channels are, of course, closed by high and opened by low concentrations of ATP. It appears that ATP-dependent K^+ channels bind ATP directly. Figure 3 shows how the P_{open} is reduced by the presence of ATP in contact with the cytoplasmic face of membrane studied under patch clamp recording. Also shown are concentration–effect relations from pancreatic B cells, heart muscle and skeletal muscle. The range of binding constant varies from around 15 to 500 μ*M*, depending upon the cell type and experimental

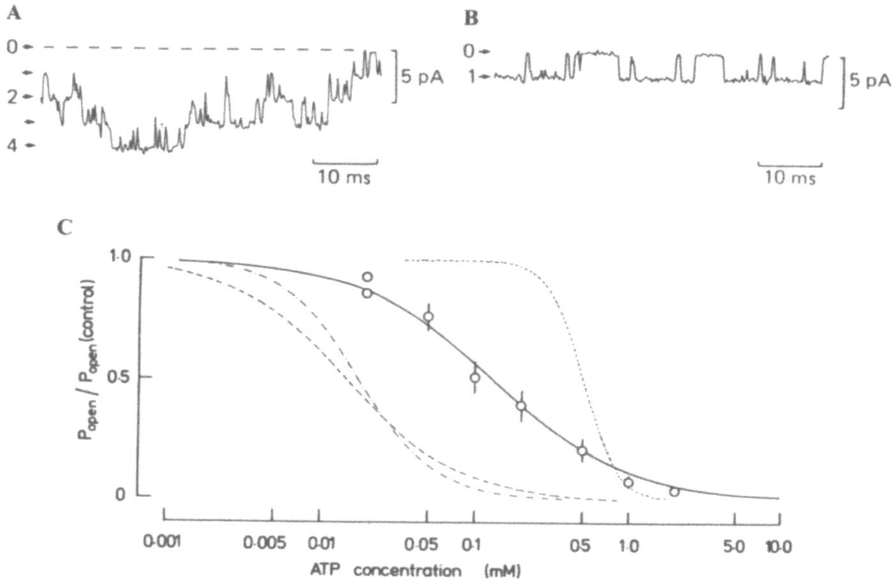

Fig. 3 A, B. Records of unitary currents from a membrane patch from frog skeletal muscle, with $[K^+]_o = 60$ m*M* and $[K^+]_i = 120$ m*M*, and with a membrane potential of -62 mV. Records in **A** are in the absence of ATP; those in **B** are in the presence of 0.5 m*M* ATP. The numbers at the left of each set of records indicate the numbers of open channels. Records filtered at 2.4 kHz and digitised at 10 kHz. Temperature, approximately 20 °C. From Spruce et al. (1987), by permission. **C** Relations between channel P_{open} and internal ATP concentration. The symbols are from results on excised inside-out membrane patches from frog skeletal muscle, replotted from Spruce et al. (1987). *Solid line* fitting the points is drawn assuming a dissociation constant of 0.135 m*M* and a Hill coefficient of 1. *Dashed line at left* indicates the results of Cook and Hales (1984) on pancreatic B cells. They used a dissociation constant of 0.015 m*M* and a Hill coefficient of 1.18. *Dashed and dotted line* indicates the fit to experimental data from excised patches of heart muscle used by Ohno-Shosaku et al. (1987), a dissociation constant of 0.018 m*M* and a Hill coefficient of 1.74. *Right-hand dashed line* gives the results of Kakei et al. (1985) in their 'open-cell' preparation of heart muscle, for which the dissociation constant was 0.5 m*M* and the Hill coefficient was greater than 3. Curve for a Hill coefficient of 4 is shown. In all cases the resting physiological intracellular ATP concentration predicts a very low or negligible P_{open} for ATP-dependent K$^+$ channels

conditions. Higher concentrations of ATP (in the millimolar range) appear to be required to close neuronal channels (Ashford et al. 1988).

As is now well-known, ATP is bound but not split in producing its pore closing effect, and non-hydrolysable analogues exert exactly the same effect as ATP itself (Noma 1983; Cook and Hales 1984; Spruce et al. 1985). Thus, neither binding nor unbinding appear to depend on hydrolysis of the nucleotide.

The stoichiometry varies from tissue to tissue. In skeletal muscle each channel may bind a single ATP; in B cells the reported Hill coefficient for binding varies between 1 (Ashcroft and Kakei 1989) and 1.18 (Cook and Hales 1984); and in heart muscle estimates vary between 1.74 (Ohno-Shosaku et al. 1987) and 3–4 (Noma 1983; Kakei et al. 1985).

Mg^{2+} Is not a Requirement for ATP Binding

Part of the evidence that ATP is not split in producing channel closure or in unbinding from the channel comes from the finding that the effect persists in the absence of Mg^{2+} (Noma 1983; Cook and Hales 1984; Spruce et al. 1985, 1987). In skeletal muscle Spruce and others (1987) reported that the effect of ATP was indistinguishable in the presence or absence of Mg^{2+}. Subsequently (Findlay 1988) it was reported that in heart, ATP had a slightly greater channel closing effect in the presence of Mg^{2+}, though others have also reported that the affinity of the channel for ATP appears independent of internal Mg^{2+} (e.g. Lederer and Nichols 1989).

The result is to some extent surprising since the form in which ATP is present in solution ($ATPMg_2$, $ATPMg^{2-}$, $ATPH^{3-}$, ATP^{4-}, etc.) clearly depends partly on the concentration of Mg^{2+} present (Table 1). The result, however, shows that Mg^{2+} is not a requirement for the binding of ATP. Furthermore, in these tissues at least, the form in which ATP exists in the internal solution does not seem to alter binding.

In pancreatic B cells, however, the apparent affinity is altered by the presence of Mg^{2+}, and adding Mg^{2+} at a constant ATP concentration increases channel P_{open} (Fig. 4). This result has generally been taken to indicate that ATP^{4-}, whose concentration is greater in the absence of the cation, is the main form that binds to the channel in pancreatic B cells. There is now good evidence to suppose that this is indeed the case. Ashcroft and Kakei (1989) have shown that in the absence of Mg^{2+}, the apparent dissociation constant

Table 1. State of ATP in the presence and absence of Mg^{2+}

Total Mg	zero	2 m*M*
Total ATP	10^{-3} *M*	10^{-3} *M*
ATP^{4-}	0.407×10^{-3} *M*	0.637×10^{-4} *M*
$ATPH^{3-}$	0.107×10^{-3} *M*	0.168×10^{-4} *M*
$ATPH_2^{2-}$	0.999×10^{-7} *M*	0.156×10^{-7} *M*
$ATPMg^{2-}$	–	0.805×10^{-3} *M*
$ATPMg_2$	–	0.355×10^{-4} *M*
$ATPK^{3-}$	0.486×10^{-3} *M*	0.764×10^{-4} *M*
$ATPMgH^{-}$	–	0.232×10^{-5} *M*

The values shown are for internal solutions whose composition is that used by Spruce et al. (1987), that is (in m*M*): KCl, 120; K_2EGTA, 1; TRIS maleate buffer (pH 7.2), 5; and $MgCl_2$, 0 or 2. Dissociation constants used are from Martell and Smith (1974).

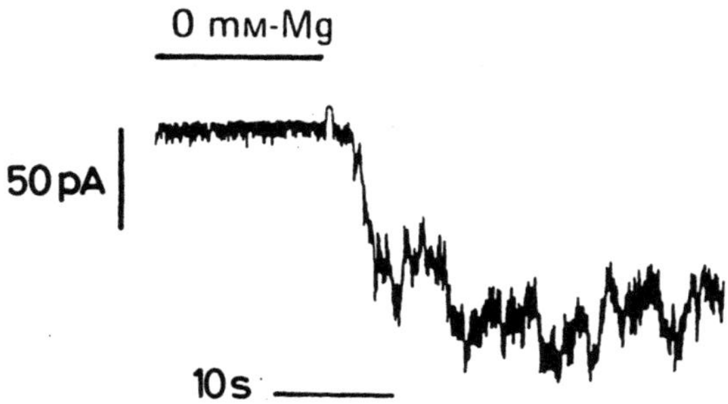

Fig. 4. Effect on ATP-dependent K$^+$ channels of pancreatic B cells of adding 5 mM Mg^{2+} to the internal face of a membrane patch already exposed to 0.1 mM ATP. The presence of Mg^{2+} reduces the channel closing ability of ATP. $[K^+]_o$, $[K^+]_i = 140$ mM, membrane potential = −60 mV. Reproduced from Ashcroft and Kakei (1987), *Journal of Physiology*, by permission

is 4 μM while in the presence of 2 mM Mg^{2+} the apparent dissociation constant rises to 26 μM. If it is assumed that it is ATP^{4-} or $ATPH^{3-}$ that binds and the concentrations of both are computed for solutions with and without Mg^{2+}, the dissociation constant is 2 μM, a value independent of the presence of Mg^{2+}. Additional experiments with analogues of ATP are consistent with the hypothesis that ionised forms of ATP bind in this cell type. Rather surprisingly, non-hydrolysable analogues of ATP (AMP-PNP or ATP-γ-S) are more effective when Mg^{2+} is present in B cells (Ashcroft and Kakei 1989). No studies appear to have been made on this point in channels in neurones or in smooth muscle.

Effect of the ATP/ADP Ratio

The control of the K$^+$ channel through binding of ATP^{4-} (or $ATPH^{3-}$) in pancreatic B cells still does not resolve the problem of the high resting physiological concentration of ATP and the high affinity of the channel for the nucleotide. While P_{open} will not be zero when the bulk cellular ATP concentration is around 5 mM, it will still be close to zero (less than 0.01 from the results of Ashcroft and Kakei 1989). If B cells have a sufficiently high resting membrane resistance, however, the high density of channels (8 μm^{-2}, Misler et al. 1986) will allow them to dominate resting membrane permeability even with an extremely low P_{open} yet also allow a reduction in P_{open} to lead to depolarisation. Such is effectively the 'spare channel' hypothesis of Cook et al. (1988). That is not true of skeletal or heart muscle, where the resting membrane resistance is around 2000–4000 ohm cm^2. In these tissues, ATP-dependent K$^+$ channels may be expected to make little or no contribution to the resting potential at resting ATP concentrations, and the postulated functions must envisage the channels being opened to alter membrane properties.

In all cell types, there has been a search for other factors that contribute to channel regulation. There is evidence in pancreatic B cells (Misler et al. 1986; Dunne and Petersen 1986a; Kakei et al. 1986) and in heart (Findlay 1988; Lederer and Nichols 1989) that the ratio of ATP to ADP is an important additional controlling mechanism. Further, it appears that Mg^{2+} is an essential co-factor in producing the effect (Findlay 1988; Lederer and Nichols 1989). In heart muscle, for example, if Mg^{2+} is absent, ADP increases the likelihood that channels are closed and has little impact on channel closure by ATP or rather increases it slightly (Findlay 1988). On the other hand, if Mg^{2+} is present in physiological concentrations, addition of ADP to the cytoplasmic face of a membrane patch reopens channels closed by ATP binding (Findlay 1988). Lederer and Nichols (1989), who describe essentially the same process, propose an allosteric effect between two nucleotide binding sites on the channel, whereby binding of ATP to one site enhances while binding of MgADP to the same site reduces the affinity of a second site for ATP. It appears that no such effect occurs in skeletal muscle since ADP has no significant effect on the P_{open} of channels closed by ATP whether Mg^{2+} is absent (Spruce et al. 1987) or present (N. W. Davies and P. R. Stanfield, unpublished observations).

In the insulin-secreting cell line RINm5F, it has been proposed that competition occurs between ATP^{4-} and ADP^{3-} (Dunne et al. 1988).

Effects of Other Nucleotides

As well as ATP and ADP, other nucleotides also affect the open state probability of ATP-dependent K^+ channels. As a first approximation, most have a similar action to ATP itself, though much higher concentrations are required to cause channel closure. This action is true in cardiac and skeletal muscle, for example, of GTP and other purine and pyrimidine nucleotides (Kakei et al., 1985; Spruce et al. 1987). Under suitable conditions, particularly in the presence of Mg^{2+}, and in certain tissues, GDP and GTP may enhance channel opening (Dunne and Petersen 1986b). The action of GDP in heart muscle is rather like that of ADP, competing for a site to reduce the affinity of a second site for ATP (Lederer and Nichols 1989). There is little action of low concentrations of GTP in heart muscle (Lederer and Nichols 1989), but in insulinoma RINm5F cells, GTP enhances the probability of channel opening rather like GDP (Dunne and Petersen 1986). As with the action of GDP, concentrations of internal Mg^{2+} around 1 m*M* seem to be a requirement for this action of GTP (Findlay 1987b). This action might be related to that of GDP in heart, reducing ATP affinity through a competitive and/or allosteric action, though it is also reminiscent of the activation of a G-protein. The possibility that these channels may be activated through activation of a G-protein is attractive now that it has been established that these channels are modulated by peptides, etc. (see above). However, the effect of galanin in insulin-secreting cell lines is insensitive to pertussis toxin and persists even in the absence of internal GTP or GTP analogues (De Weille et al. 1988).

In the Presence of Mg^{2+}, ATP Protects ATP-Dependent K^+ Channels from Rundown

Most authors who have described ATP-dependent K^+ channels have reported that the channel activity of excised membrane patches decreases with time (Kakei et al. 1985; Findlay and Dunne 1986; Spruce et al. 1987; Ohno-Shosaku et al. 1987), though the rate at which this occurs varies from tissue to tissue. This rundown can, however, be reduced in rate by a relatively brief exposure of the channel to ATP (Findlay and Dunne 1986; Ohno-Shosaku et al. 1987), which of course temporarily closes channels but results in an increase in the number of channels open once the ATP is again removed experimentally. Unlike the channel-closing effect of the nucleotide, this effect on rundown requires internal Mg^{2+} (Fig. 5; Ohno-Shosaku et al. 1987; Findlay 1987b) and is probably associated with channel rephosphorylation. Other regulating effects through phosphorylation, for example by protein kinase C (Dunne and Petersen 1987), may also occur (see however Plant 1989, who has demonstrated non-specific effects of synthetic cell permeable diacylglycerols on ion channels in B cells).

Internal Mg^{2+} May Reduce Channel P_{open}

In addition to being a co-factor in the regulation of ATP-dependent K^+ channels by intracellular nucleotides, Mg^{2+} appears to have a direct effect on the open state probability of the channels and on the efflux of K^+ through them. In pancreatic B cells it seems clear that, in the absence of ATP, removal of Mg^{2+} from the intracellular solution increases channel P_{open}, while addition of Mg^{2+} reduces it (Ashcroft and Kakei 1987; Findlay 1987a, b). Ashcroft and Kakei (1989) show that the effect is consistent with Mg^{2+} binding to the

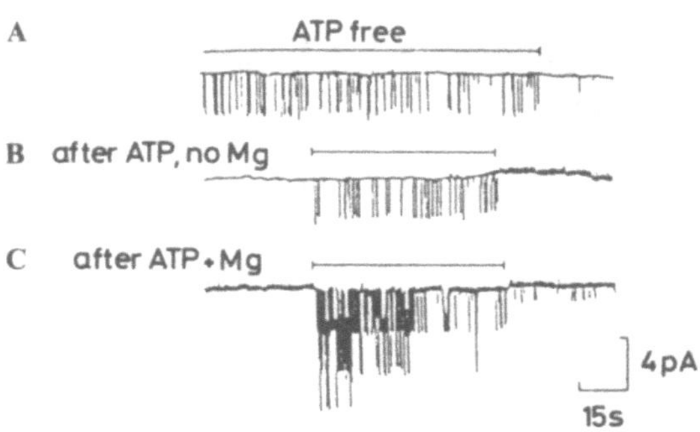

Fig. 5A–C. Role of Mg^{2+} in aiding ATP in the reversal of rundown of channels. **A** Record from a membrane patch exposed to ATP-free solution for several minutes. **B** Activation by removal of ATP applied at 0.3 m*M* in the absence of Mg^{2+}. **C** Activation by removal of ATP (0.3 m*M*) applied with 1 m*M* Mg^{2+}. Note the higher P_{open} in **C** compared with **B**. Reproduced from Ohno-Shosaku et al. (1987), *Pflügers Archiv*, by permission

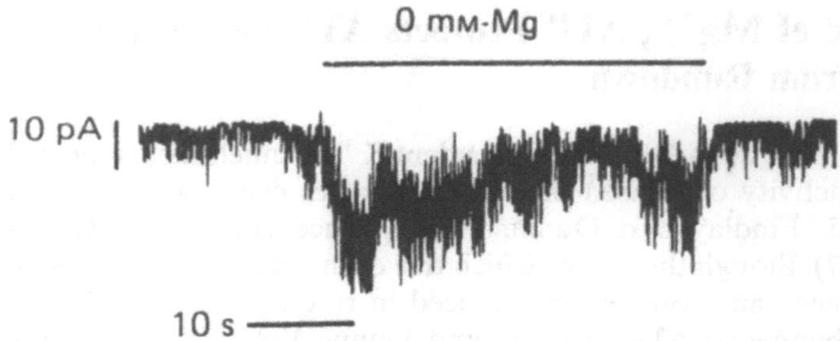

Fig. 6. Effect of removing internal Mg^{2+} (5 m*M* when present) in the absence of ATP on pancreatic B cells. $[K^+]_o$ and $[K^+]_i = 140$ m*M*, membrane potential -60 mV. Reproduced from Ashcroft and Kakei (1987), *Journal of Physiology*, by permission

channel with a dissociation constant of 5 m*M* and a Hill coefficient of 1.9, indicating co-operativity. The effect at physiological Mg^{2+} concentration may therefore be expected to be small. In skeletal muscle, we find no influence of Mg^{2+} on P_{open} (N. W. Davies and P. R. Stanfield, unpublished work).

The mechanism of action of Mg^{2+} is not fully understood, though it is an action that is shared with other divalent cations. Noma (1983) reported that 10^{-5} *M* Ca^{2+} reduced channel P_{open} in heart muscle, and Findlay (1987a, b) has confirmed this observation for both heart and insulin-secreting cells. Ca^{2+} is effective at considerably lower concentrations than Mg^{2+}, as are Ba^{2+} and Sr^{2+}. Although the effect on P_{open} is common to a number of divalents, it seems unlikely to be an effect on internal surface potential and membrane voltage field, since channel P_{open} is relatively little affected by membrane potential. As with Mg^{2+}, we find no significant effect of Ca^{2+} on channel P_{open} in skeletal muscle at concentrations up 10^{-5} *M* (Spruce et al. 1987).

Internal Mg^{2+} Blocks ATP-Dependent K^+ Channels in a Voltage-Dependent Manner

A number of ion channels are known to be the subject of voltage-dependent blockage by impermeant intracellular ions, generally Mg^{2+} or Na^+. Blockage by Mg^{2+} was first demonstrated in NMDA receptors by Nowak et al. (1984; see Johnson and Ascher, this volume) and established as a mechanism of producing inward rectification of K^+ channels in ATP-dependent K^+ channels of heart by Horie et al. (1987). Such a mechanism is now known to occur in K^+ channels of heart activated by muscarinic acetylcholine receptors (Horie and Irisawa 1987), in inward rectifiers (Matsuda et al. 1987; Vandenburg 1987; see Chap. 4, p. 51, this volume), in neuronal-type acetylcholine receptors (Cachelin and Neuhaus 1989) and probably in channels carrying A-current (Forsythe and Stanfield 1989).

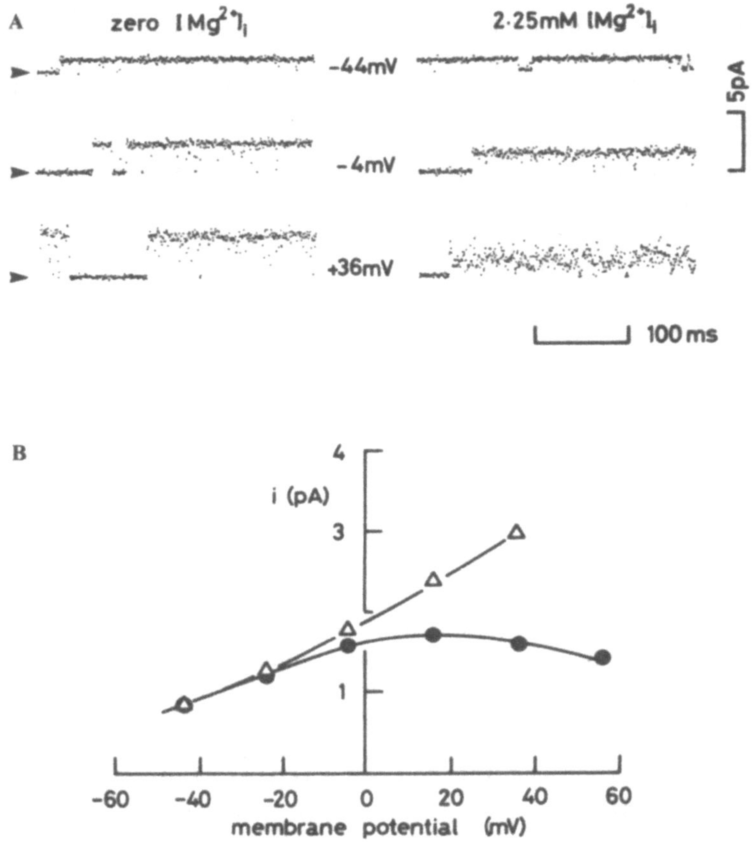

Fig. 7 A, B. Internal Mg^{2+} as a blocking ion of ATP-dependent K^+ channels of skeletal muscle. **A** Records of unitary currents with $[K^+]_o = 2.5$ m*M*, $[K^+]_i = 120$ m*M*, in the presence (*right*) and absence (*left*) of 2.25 m*M* Mg^{2+}. Records filtered at 1 kHz, digitised at 2 kHz. Courtesy of Dr JM Quayle. **B** Current–voltage relations in absence (△) and presence (●) of Mg^{2+}. Line to the points in Mg^{2+} is drawn with $K(0) = 11.4$ m*M* and $z' = 0.98$ (see text)

Several authors besides Horie et al. (1987) have shown a voltage-dependent block of ATP-dependent K^+ channels by internal Mg^{2+}. It has also been demonstrated in heart muscle by Findlay (1987 a) and in the insulinoma cell line RINm5F by Findlay (1987 b) and by Ciani and Ribalet (1988). Figure 7 shows the phenomenon in ATP-dependent K^+ channels of skeletal muscle, where the voltage-dependent block was demonstrated in my laboratory by John Quayle (1988). The records in Fig. 7 are from an excised inside-out patch whose cytoplasmic face was in contact with a solution containing either zero or 2.25 m*M* Mg^{2+}. Little blockage occurs at negative membrane potentials, but the effect of Mg^{2+} becomes noticeable as the potential approaches 0 mV. At positive potentials, blockage is obvious and is associated with a striking increase in noise in the open current level as Mg^{2+} rapidly enters and leaves the channel. In heart musle, this blockage by internal cations is presumably important in limiting K^+ permeability in the voltage range of the plateau of the cardiac action potential. In skeletal muscle the function of the block is less clear. In skeletal muscle, the affinity for Mg^{2+} is lower, but the block is more steeply voltage-dependent. Mg^{2+} blockage may occur primarily in K^+ channels that express their physiological function in the region of the resting potential and at slightly more positive potentials, rather than in channels that operate to repolarise action potentials.

In general, the blockage of a channel by an impermeant ion such as Mg^{2+} can be characterised by expressing the fractional block ($y = 1 - i_{Mg}/i_{control}$; where i_{Mg} is the unitary current in the presence of the blocking ion and $i_{control}$ is that in its absence) in terms of the concentration of the blocking ion and a dissociation constant:

$$y = \{1 + [Mg^{2+}]/K(V)\}^{-1}$$

where $K(V)$ is the voltage-dependent binding constant, falling under depolarisation according to:

$$K(V) = K(0)\exp(-z'\,VF/RT)$$

Here $K(0)$ is the dissociation constant at 0 mV membrane potential and z' is the so-called effective valency of the blocking ion, equal to the actual valency multiplied by the fraction of the membrane voltage field acting on the ion as it moves between the internal solution and its blocking site. The block in frog skeletal muscle may under physiological conditions, 2.5 mM $[K^+]_o$, be fitted with $K(0) = 11.4$ mM and $z' = 0.98$ (Fig. 7). In 60 mM $[K^+]_o$, $K(0)$ is increased to approximately 24 mM as competition between the permeant K^+ and the blocking Mg^{2+} reduces the affinity of the ion channel for the divalent cation.

From the more detailed studies in mammalian heart muscle by Horie et al. (1987), the value for the dissociation constant $K(0)$ was 2.0 mM at the physiological $[K^+]_o$ of 5.4 mM, and the effective valency, z', was approximately 0.6. Again, increasing $[K^+]_o$ reduced affinity, $K(0)$ increasing about fourfold when $[K^+]_o$ was increased from 5.4 to 140 mM. Since the block by Mg^{2+} is very rapid, and blocking and unblocking events are incompletely resolved as discrete events with the low pass filtering used in single channel recording, Horie et al. (1987) used noise analysis to investigate the transition rates between blocked and unblocked channels, considering a model in which channels exist in the following states:

$$\text{closed} \underset{}{\overset{\text{ATP}}{\rightleftharpoons}} \text{closed} \overset{\text{ATP}}{\rightleftharpoons} \text{open} \underset{k_{-1}}{\overset{Mg^{2+}\ k_1}{\rightleftharpoons}} \text{blocked}$$

though there is as yet no evidence that closed channels cannot be blocked by Mg^{2+} as this linear scheme implies (and transitions between closed states and between closed and open do not necessarily involve the unbinding of ATP; see Fig. 1). The blocking transition rate will be $[Mg^{2+}]_i\,k_1$, and the unblocking rate k_{-1}. Noise analysis suggests that a single Lorentzian will fit the power spectrum for the excess current noise associated with Mg^{2+} block. This suggests that blockage is indeed a first-order process as the above scheme implies. The corner frequency, f_c, of the Lorentzian function will, if channel gating does not interfere, approximate $\{[Mg^{2+}]_i \cdot k_1 + k_{-1}\}/2\pi$, and the rate constants may be evaluated since $K(V)$, the dissociation constant, is also known. The corner frequencies were greater than 1 kHz, and the rate constants have values with 5.4 mM $[K^+]_o$ and at 0 mV in the order of $10^7\ M^{-1}\ s^{-1}$ for k_1 and $10^4\ s^{-1}$ for k_{-2}. The blocking rate k_1 is more voltage dependent than the unblocking rate k_{-1}, as often appears to be the case for voltage-dependent ion blocking mechanisms (see for example, Quayle et al. 1988). It was also shown that k_{-1} is increased by increasing $[K^+]_o$ as might be expected if the blocking ion is flushed out of the channel by inward movement of K^+.

In insulin-secreting RINm5F cells, the dissociation constant for Mg^{2+} was 34 mM in symmetrical 140 mM K^+ solutions, and the effective valency was 0.7 (Ciani and Ribalet 1988).

Other Intracellular Factors in the Control of ATP-dependent K^+ Channels

Although internal Mg^{2+} apears to modulate the open state probability of ATP-dependent K^+ channels in certain cells (heart and islets of Langerhans) and reduces the efflux of K^+ at positive voltages, it is not the only intracellular ion affecting channel function. Both Na^+ and H^+ appear able to block ATP-dependent K^+ channels from the internal solution, and changes of internal pH alter both channel kinetic behaviour and the response to ATP.

Na^+ Block

In one of the earliest accounts of ATP-dependent K^+ channels (Cook and Hales 1984), a voltage-dependent block by Na^+ was believed to be the cause of the inward rectification of the unitary current voltage relation. For ATP-dependent K^+ channels, this may still be at least as important as that by Mg^{2+}, certainly for the channels in skeletal muscle. In skeletal muscle the blockage by Na^+ can be fitted with $K(0) = 22$ mM (2.5 mM external K^+) and $z' = 0.74$ (Quayle and Stanfield 1989). In heart (Horie et al. 1987), $K(0)$ is approximately 25 mM with $[K^+]_o = 5.4$ mM (15 mM at +40 mV) and $z' = 0.35$. In this case a Hill coefficient of 2 was found. In the insulin-secreting cell line RINm5F, a site close to the intracellular solution binds Na^+ with a dissociation constant of 46 mM (Ciani and Ribalet 1988). In skeletal muscle, Na^+ permeates again at very positive voltages as the membrane voltage field drives the blocking ion through the channel selectivity filter (Quayle and Stanfield 1989). In all cases, Na^+ blockage is extremely rapid, reducing single channel amplitude without increasing noise in the open level.

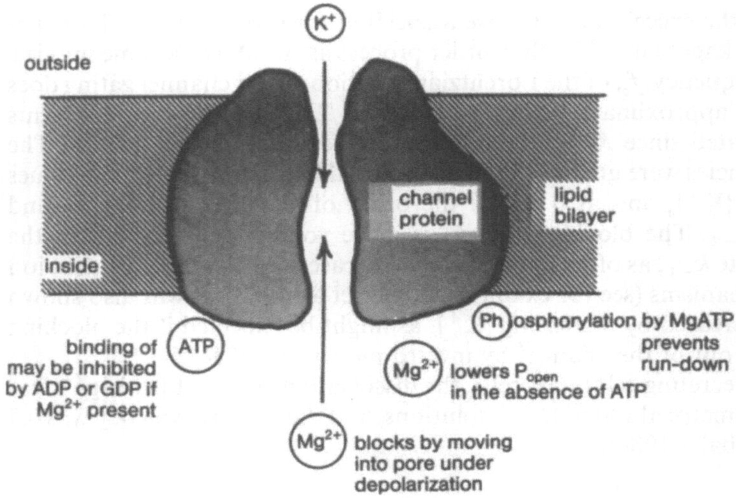

Fig. 8. Summary of the roles played by Mg^{2+} in the regulation of ATP-dependent K^+ channels. From left to right across the bottom of the figure, the roles are as follows. (i) Binding of ATP does not require Mg^{2+}, and in heart and skeletal muscle ATP binding is unaltered whether Mg^{2+} is present or not. In B cells Mg^{2+} reduces binding by reducing the availability of ATP^{4-}. Effect of ADP in counteracting channel closure by ATP does require the cation (B cells, heart muscle, no effect in skeletal muscle). (ii) Mg^{2+} produces a voltage-dependent blockage of the channel, as it is driven into the channel by depolarisation. External K^+ counteracts the block, increasing the transition rate for Mg^{2+} leaving the channel. (iii) Mg^{2+} lowers channel P_{open} in heart and B cells. This effect is slight at physiological Mg^{2+} concentrations (Ashcroft and Kakei 1989). (iv) Mg^{2+} is a co-factor in channel phosphorylation. It is likely that phosphorylation at one site protects that channel from rundown, which occurs experimentally in membrane patches excised for single channel recording. Since the channels are also activated by peptides, it seems likely that the channel will be modulated by other second messengers and/or G-protein activation. Mg^{2+} is likely to be a co-factor in such steps

Intracellular Protons

Acidification of the internal solution also results in a reduction of single channel amplitude, which is little dependent on voltage (Davies et al. 1989 b). In the absence of ATP, acidification results in a substantial increase in the duration of channel openings, with little effect on open state probability (Davies et al. 1989 b). In the presence of ATP, however, open state probability is increased by acidification (Spruce 1986; Davies 1990).

Summary

Whatever the effects of these other intracellular cations, Mg^{2+} is clearly emerging as a co-factor in ion channel function. For the ATP-dependent K^+

channels of excitable membranes, Mg^{2+} has effects on the binding of nucleotides, influencing channel P_{open}; has direct effects on channel P_{open} at least in some cells; and works as a channel blocker contributing to a reduction in K^+ efflux through channels at positive voltages. Some of these effects are summarised in Fig. 8.

References

Ashcroft FM, Kakei M (1987) Effects of internal Mg^{2+} on ATP-sensitive K-channels in isolated rat pancreatic B-cells. J Physiol (Lond) 390:72P

Ashcroft FM, Kakei M (1989) ATP-sensitive K^+ channels in rat pancreatic B-cells: modulation by ATP and Mg^{2+} ions. J Physiol (Lond) 416:349–367

Ashcroft FM, Harrison DE, Ashcroft SJH (1984) Glucose induces closure of single potassium channels in isolated rat pancreatic B-cells. Nature 312:446–448

Ashford MLJ, Sturgess NC, Trout NJ, Gardner MJ, Hales CN (1988) Adenosine-5′-triphosphate sensitive ion channels in neonatal rat cultured central neurones. Pflügers Arch 412:297–304

Ashford MLJ, Boden PR, Treherne JM (1989) Glucose-induced excitation of rat hypothalamic neurones in vitro is mediated by ATP-sensitive K^+ channels. J Physiol (Lond) 415:31P

Ben Ari Y, Krnjevic K (1989) The ATP-sensitive K^+ channel opener diazoxide reduces anoxic depolarization in rat CA3 hippocampal neurones in isolated slices. J Physiol (Lond) 418:192P

Cachelin AB, Neuhaus R (1989) The effect of Mg^{2+} on currents through single acetylcholine channels of rat-cultured phaeochromocytoma cells. J Physiol (Lond) 418:38P

Carlson FD, Siger A (1960) The mechanochemistry of muscular contraction. I. The isometric twitch. J Gen Physiol 43:33–60

Ciani S, Ribalet B (1988) Ion permeation and rectification in ATP-sensitive channels from insulin-secreting cells (RINm5F): effects of K^+, Na^+ and Mg^{2+}. J Membr Biol 103:171–180

Clay JR, Szuts EZ (1988) Modulation of potassium channel inactivation in squid axons by ATP. Biol Bull 175:315

Cook DL, Hales CN (1984) Intracellular ATP directly blocks K^+ channels in pancreatic B-cells. Nature 311:271–273

Cook DL, Satin LS, Ashford MLJ, Hales CN (1988) ATP-sensitive K^+ channels in pancreatic B-cells. Spare-channel hypothesis. Diabetes 37:495–498

Davies NW (1990) ATP-sensitive K channels in skeletal muscle are modulated by intracellular protons. Nature 343:375–377

Davies NW, Huang Y, Nelson MT, Quayle JM, Standen NB (1989a) ATP-sensitive potassium channels recorded from isolated arterial smooth muscle cells of rabbit and rat. J Physiol (Lond) 415:29P

Davies NW, Standen NB, Stanfield PR (1989b) The kinetic behaviour of K_{ATP} channels in isolated membrane patches of frog skeletal muscle is regulated by internal pH. J Physiol (Lond) 418:188P

De Weille J, Schmid-Antomarchi H, Fosset M, Lazdunski M (1988) ATP-sensitive K^+ channels that are blocked by hypoglycaemia-inducing sulfonylureas in insulin-secreting cells are activated by galanin, a hyperglycaemia-inducing hormone. Proc Natl Acad Sci USA 85:1312–1316

Dunne MJ, Petersen OH (1986a) Intracellular ADP activates K^+ channels that are inhibited by ATP in an insulin-secreting cell line. FEBS Lett 208:59–62

Dunne MJ, Petersen OH (1986b) GTP and ADP activate ATP-inhibited K channels in the insulin-secreting cell line RINm5F. J Physiol (Lond) 381:67P

Dunne MJ, Petersen OH (1987) Phorbol ester and cell-permeable diacylglycerol evoke closure of ATP-sensitive K^+ channels in a cultured insulin-secreting cell line. J Physiol (Lond) 390:73P

Dunne MJ, West-Jordan JA, Abraham RJ, Edwards RHT, Petersen OH (1988) The gating of nucleotide-sensitive K^+ channels in insulin-secreting cells can be modulated by changes in the ratio ATP^{4-}/ADP^{3-} and by non-hydrolysable analogues of both ATP and ADP. J Membr Biol 104:165–177

Findlay I (1987a) ATP-sensitive K^+ channels in rat ventricular myocytes are blocked and inactivated by internal divalent cations. Pflügers Arch 410:313–320

Findlay I (1987b) The effects of magnesium ions upon adenosine triphosphate-sensitive potassium channels in a rat insulin-secreting cell line. J Physiol (Lond) 391:611–629

Findlay I (1988) Effects of ADP upon the ATP-sensitive K^+ channel in rat ventricular myocytes. J Membr Biol 101:83–92

Findlay I, Dunne MJ (1986) ATP maintains ATP-inhibited K^+ channels in an operational state. Pflügers Arch 407:238–240

Fink R, Hase S, Lüttgau HC, Wettwer E (1983) The effect of cellular energy reserves and internal calcium ions of the potassium conductance in skeletal muscle of the frog. J Physiol (Lond) 336:211–228

Fink R, Lüttgau HC (1976) An evaluation of the membrane constants and the potassium conductance in metabolically exhausted muscle fibres. J Physiol (Lond) 263:215–238

Forsythe ID, Stanfield PR (1989) Single channel recording of A-current in neurones from the rat locus coeruleus, grown in dissociated cell culture. J Physiol (Lond) 417:85P

Fujiwara N, Higashi H, Shimoji K, Yoshimura M (1987) Effects of hypoxia on rat hippocampal neurones in vitro. J Physiol (Lond) 384:131–151

Godfraind JM, Krnjevic K, Pumain R (1970) Unexpected features of the action of dinitrophenol on cortical neurones. Nature 228:562–564

Horie M, Irisawa H (1987) Rectification of muscarinic K^+ current by magnesium ion in guinea-pig atrial cells. Am J Physiol 253:H210–H214

Horie M, Irisawa H, Noma A (1987) Voltage-dependent magnesium block of adenosine-triphosphate-sensitive potassium channel in guinea-pig ventricular cells. J Physiol (Lond) 387:251–272

Kakei M, Noma A, Shibasaki T (1985) Properties of adenosine triphosphate regulated potassium channels in guinea-pig ventricular cells. J Physiol (Lond) 363:441–462

Kakei M, Kelly RP, Ashcroft SJH, Ashcroft FM (1986) The ATP-sensitivity of K^+ channels in rat pancreatic B-cells is modulated by ADP. FEBS Lett 208:63–66

Krnjevic K, Lisiewicz A (1972) Injections of calcium ions into spinal motoneurones. J Physiol (Lond) 225:363–390

Lederer WJ, Nichols CG (1989) ADP and GDP modulate the ATP-sensitivity of K_{atp} channels in isolated rat ventricular cell membranes. J Physiol (Lond) 415:28P

Martell AF, Smith RM (1974) Critical stability constants. Plenum, New York

Matsuda H, Saigusa A, Irisawa H (1987). Ohmic conductance through the inwardly rectifying K channels and blocking by internal Mg^{2+}. Nature 325:156–159

Meech RW, Strumwasser F (1970) Intracellular calcium injection activates potassium conductance in *Aplysia* nerve cells. Fed Proc 29:834

Misler S, Falke LC, Gillis K, McDaniel ML (1986) A metabolite-regulated potassium channel in rat pancreatic B-cells. Proc Natl Acad Sci USA 83:7119–7123

Mourre C, Ben Ari Y, Bernardi H, Fosset M, Lazdunski M (1989) Antidiabetic sulfonylureas: localization of binding sites in the brain and effects on the hyperpolarization induced by anoxia in hippocampal slices. Brain Res 486:159–164

Noma A (1983) ATP-regulated K^+ channels in cardiac muscle. Nature 305:147–148

Nowak L, Bregestovski P, Ascher P, Herbet A, Prochiantz A (1984) Magnesium gates glutamate-activated channels in mouse central neurones. Nature 307:462–465

Ohno-Shosaku T, Zünkler BJ, Trube G (1987) Dual effects of ATP on K^+ currents of mouse pancreatic B-cells. Pflügers Arch 408:133–138

Ono T, Nishino H, Fukuda M, Sasaki K, Muramoto K-I, Oomura Y (1982) Glucose responsive neurones in rat ventromedial hypothalamic slices in vitro. Brain Res 232:494–499
Paulais M, Toulon J (1989) A cation channel in the thick asending limb of Henle's loop of the mouse kidney: inhibition by adenine nucleotides. J Physiol (Lond) 413:315–327
Plant TD (1989) Effects of 1,2-diocytanoyl-sn-glycerol (DiC_8) on ionic currents in cultured mouse pancreatic B-cells. J Physiol (Lond) 418:21P
Quayle JM (1988) Kinetics and block of the ATP-sensitive potassium channel of frog skeletal muscle. PhD thesis, University of Leicester
Quayle JM, Stanfield PR (1989) Block of the ATP-sensitive potassium channel of frog skeletal muscle of internal sodium ions. J Physiol (Lond) 410:85P
Quayle JM, Standen NB, Stanfield PR (1988) The voltage-dependent block of ATP-sensitive potassium channels of frog skeletal muscle by caesium and barium ions. J Physiol (Lond) 405:677–697
Rorsman P, Trube G (1985) Glucose-dependent K^+-channels in pancreatic B-cells are regulated by intracelluar ATP. Pflügers Arch 405:305–309
Smith JS, Coronado R, Meissner G (1985) Sarcoplasmic reticulum contains adenine nucleotide-activated calcium channels. Nature 316:446–449
Spruce AE (1986) Potassium conductances of skeletal muscle investigated using single channel recording. PhD thesis, University of Leicester
Spruce AE, Standen NB, Stanfield PR (1985) Voltage-dependent, ATP-sensitive potassium channels of skeletal muscle membrane. Nature 316:446–449
Spruce AE, Standen NB, Stanfield PR (1987) Studies of the unitary properties of adenosine-5′-triphosphate-regulated potassium channels of frog skeletal muscle. J Physiol (Lond) 383:213–236
Standen NB, Quayle JM, Davies NW, Brayden JE, Huang Y, Nelson MT (1989) ATP-sensitive K channels in arterial smooth muscle and the action of hyperpolarizing vasodilators. Science 245:177–180
Stanfield PR (1987) Nucleotides such as ATP may control the activity of ion channels. Trends Neurosci 10:335–339
Sturgess NC, Hales CN, Ashford MLJ (1987) Calcium and ATP regulate the activity of a non-selective cation channel in a rat insulinoma cell line. Pflügers Arch 409:607–615
Taylor SG, Weston AH (1988) Endothelium-derived hyperpolarizing factor: a new endogenous inhibitor from the vascular endothelium. Trends Pharmacol Sci 9:272–274
Trautwein W, Gottstein U, Dudel J (1954) Der Aktionsstrom der Myokardfaser im Sauerstoffmangel. Pflügers Arch 260:40–60
Vandenburg CA (1987) Inward rectification of a potassium channel in cardiac ventricular cells depends on internal magnesium ions. Proc Natl Acad Sci USA 84:2560–2564
Whittam R (1968) Control of membrane permeability to potassium in red blood cells. Nature 219:610

Voltage-Dependent Blockage of Cardiac Inwardly Rectifying K^+ Channels by Internal Mg^{2+}

Hiroko Matsuda[1]

Introduction

Up to now six kinds of cardiac potassium channels have been identified in single-channel recordings (Hamill et al. 1981): the inwardly rectifying K^+ channel (Trube et al. 1981; Kameyama et al. 1983; Sakmann and Trube 1984 a, b), the muscarinic receptor-operated K^+ channel (Sakmann et al. 1983), the adenosine 5′-triphosphate (ATP)-regulated K^+ channel (Noma 1983), the Na^+-activated K^+ channel (Kameyama et al. 1984), the Ca^{2+}-activated K^+ channel (Callewaert et al. 1986) and the delayed rectifier K^+ channel (Shibasaki 1987; Yue and Marban 1988). Most of them show inward rectification, which plays an important role in maintaining the long-lasting action potentials characteristic of cardiac cells.

The phenomenon of inward rectification was first discovered in the resting K^+ conductance of skeletal muscle close to 40 years ago (Katz 1949). The mechanism that allowed K^+ to move readily into the cell but not out remained a puzzle. Several possibilities have been considered (Stanfield 1988): (1) channels themselves rectify, because of the arrangement of energy barriers and wells that K^+ traverses as it permeates the channel (Woodbury 1971); (2) channels are gated by voltage and are closed under depolarisation; and (3) there is some intracellular blocking particle or blocking ion that prevents K^+ efflux by moving into the inner mouth of the K^+ channel and is driven intracellularly again at negative voltage by inward-moving K^+ (Armstrong 1975; Hille and Schwarz 1978; Standen and Stanfield 1978).

The idea that the blocking ion is at least partly responsible for inward rectification turns out to be correct in the heart muscle: the blocking ion producing rectification is Mg^{2+}. That intracellular Mg^{2+} blocks outward currents selectively was first observed in the ATP-regulated K^+ channel (Horie et al. 1987) and then in the inwardly rectifying K^+ channel (Matsuda et al. 1987; Vandenberg 1987; Matsuda 1988), the muscarinic receptor-operated K^+ channel (Horie and Irisawa 1987, 1989) and the Na^+-activated K^+ channel (M. Horie, personal communication). Unidirectional and voltage-dependent blockage by internal Mg^{2+} seems to be a common mechanism causing inward rectification of cardiac K^+ channels. However, sensitivity to Mg^{2+} differs

[1] Department of Physiology, Faculty of Medicine, Kyushu University 60, Fukuoka 812, Japan

among the kinds of channels. This paper will review mainly research on the Mg^{2+} blockage of the inwardly rectifying K^+ channel responsible for the resting K^+ conductance and compare it with that in other cardiac K^+ channels.

Ohmic Conductance Through the Inwardly Rectifying K^+ Channel

Internal Dialysis with Low-Divalent Cation Solution Reveals Outward Currents Through the Inwardly Rectifying K^+ Channel

The kinetics of the inwardly rectifying K^+ channel has been studied extensively in the voltage range negative to the equilibrium potential for K^+ (E_K; for review, see Hagiwara 1983). However, in spite of its physiological importance, little is known of the kinetics at voltages positive to E_K because of the difficulty

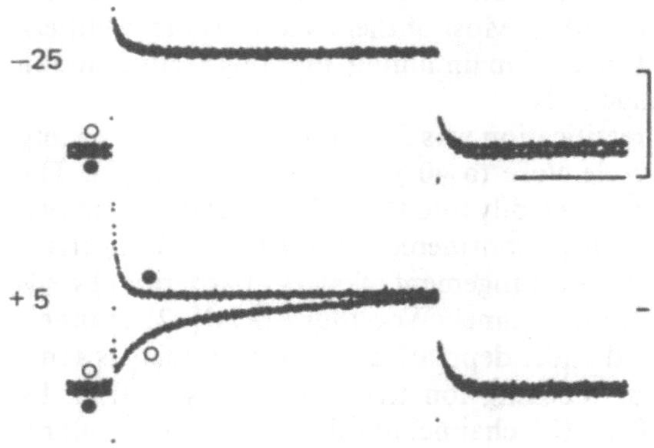

Fig. 1. Whole-cell currents obtained before (*open circles*) and after (*filled circles*) Co^{2+} application. Cell was perfused with Cs^+-rich internal solution. Membrane potential was stepped from −45 mV to the potential indicated (in mV) to the *left* of each current trace. In 5.4 m*M* K^+, Na^+-free external solution, 0.9 m*M* Co^{2+} was substituted for an equimolar amount of Ca^{2+}. *Marks to the right* of each record indicate zero current. Time calibration indicates 100 ms and current calibration represents 1 nA. (From Matsuda and Noma 1984)

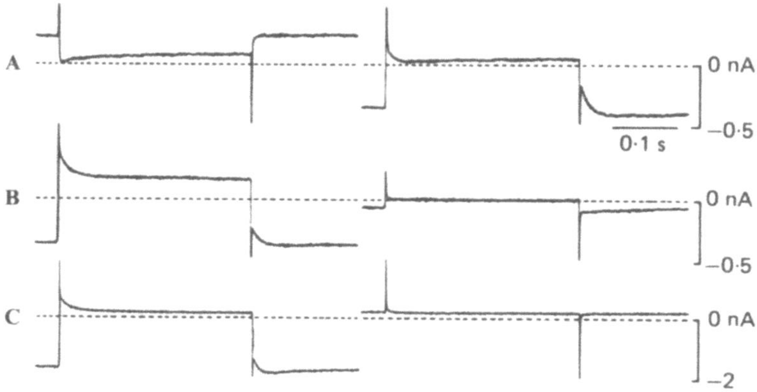

Fig. 2. A Induction of the time-dependent current by replacing the K^+-rich internal solution (*left*) with the Cs^{2+}-rich one (*right*) in Na^+-free external solution containing 5.4 m*M* K^+, 0.9 m*M* Co^{2+} and 0.9 m*M* Ca^{2+}, given 300-ms step depolarisations from −45 to −5 mV. **B** Current records before (*left*) and after (*right*) addition of 0.2 m*M* Ba^{2+} to Na^+-free external solution containing 5.4 m*M* K^+, 0.9 m*M* Co^{2+} and 0.9 m*M* Ca^{2+}, with Cs^+-rich internal solution, given 300-ms step depolarisations from −45 to −25 mV. **C** Current records with Cs^+-rich internal solution in Na^+-free external solution containing 5.4 m*M* K^+ (*left*) and 0 m*M* K^+ (*right*), given 300-ms step depolarisations from −45 to −25 mV. Temperature of solution in chamber was kept at 36°−37 °C in the experiments shown in Figs. 1 and 2. (From Matsuda and Noma 1984)

in recording the outward currents through the channel. A clue to overcoming this difficulty was provided by experiments using the intracellular perfusion technique (Matsuda and Noma 1984). Figure 1 shows current traces obtained from a single, guinea pig, ventricular cell perfused with Cs^+-rich, divalent ion-free internal solution at 5.4 m*M* external K^+. The zero current potential, which can be considered approximately equal to E_K, was around −31 mV under this condition. The membrane potential was depolarised from −45 mV to the indicated level. In the absence of a Ca^{2+} current (with a smaller test potential or after Co^{2+} application), a decaying outward current follows the positive spike of capacitive current during depolarising pulses, and the time course of current decay is accelerated with larger depolarisation. An increasing inward current is observed on repolarisation to the holding potential.

Such time-dependent currents were induced by lowering the intracellular K^+ concentration as shown in Fig. 2A, where a K^+-rich internal solution was exchanged for a Cs^+-rich solution while exposing the cell to 0.9 m*M* Co^{2+} to block Ca^{2+} currents. Either the addition of 0.2 m*M* Ba^{2+} to the external solution (Fig. 2B) or the removal of external K^+ (Fig. 2C) suppressed both the inward and outward time-dependent currents. Low concentrations of Ba^{2+} are known to suppress K^+ currents, especially currents flowing through inwardly rectifying K^+ channels (Sakmann and Trube 1984b). The removal of external K^+ also abolishes the inwardly rectifying component of the steady-state current–voltage (I–V) relationship (Isenberg and Klöckner 1982; Matsuda and Noma 1984); the concomitant disappearance of the time-dependent current provides a strong argument which postulates that the steady-state and time-dependent currents both flow in inwardly rectifying K^+ channels.

The well-known dependence of inwardly rectifying K^+ conductance on the driving force for K^+ movement across the cell membrane (e.g. Hodgkin and Horowicz 1959) provides a possible explanation for the appearance of the time-dependent K^+ current upon reducing intracellular K^+ concentration. With K^+-rich pipette solution, the holding potential of −45 mV lies some 40 mV positive to E_K. Inwardly rectifying K^+ channels should be mostly closed in the steady state, and depolarisation should lead only to further closing, so that voltage pulses between −45 and 0 mV are not expected to be associated with much change in K^+ current. However, when intracellular K^+ concentration is lowered by substitution with Cs^+, E_K is shifted to about −30 mV, and, since the K^+ driving force is then inward, inwardly rectifying K^+ channels are expected to be open. Depolarisation to voltages positive to E_K should then lead to channel closure and repolarization to channel reopening, events presumably causing changes in membrane K^+ current.

This possibility was tested in another experiment in which E_K was shifted to about −30 mV by increasing the external K^+ concentration from 5.4 m*M* to 40 m*M* with 150 m*M* K^+ internal solution (Fig. 3). The temperature of the solution in the chamber was kept at 15 °C to slow the kinetics and make the analysis easy. (It is the same in all following figures, unless otherwise indicated.) The membrane was hyperpolarised from the zero current potential by 30 mV and then depolarised to a potential more positive than the zero current potential by 20 mV. During hyperpolarisation, an inward current was activated, and following depolarisation, a decaying outward current was elicited (Fig. 3 A). Prolongation of the hyperpolarising prepulse increased the amplitude of the decaying current with a time course (Fig. 3 B) similar to the increase of the inward current during the prepulse (Fig. 3 C). This suggests that the outward current flows through the inwardly rectifying K^+ channel and that the current decrease in depolarisation reflects deactivation of the channel activated during the preceding hyperpolarisation.

Time-dependent changes in both outward and inward currents could be fitted with a single exponential function (Fig. 3 C for inward currents and Fig. 4 A for outward currents). The time constant of the inward or outward current change decreased exponentially as the voltage step from the zero current potential was increased (Fig. 4 A), and the time constant–voltage relation was bell-shaped with a peak (8–9 ms) at around E_K (Fig. 5 in Saigusa and Matsuda 1988).

The instantaneous current was obtained by extrapolating the best fitted curve to the onset of the depolarising pulse and plotted against the membrane potential together with the steady-state current (Fig. 4 B). Almost linear instantaneous I–V curves were obtained in 15 out of 21 cells. This finding suggests that the I–V relation of the inwardly rectifying K^+ channel is ohmic.

Ohmic conductance was demonstrated in single-channel current recordings as will be described below. However, some remarks should be made on the

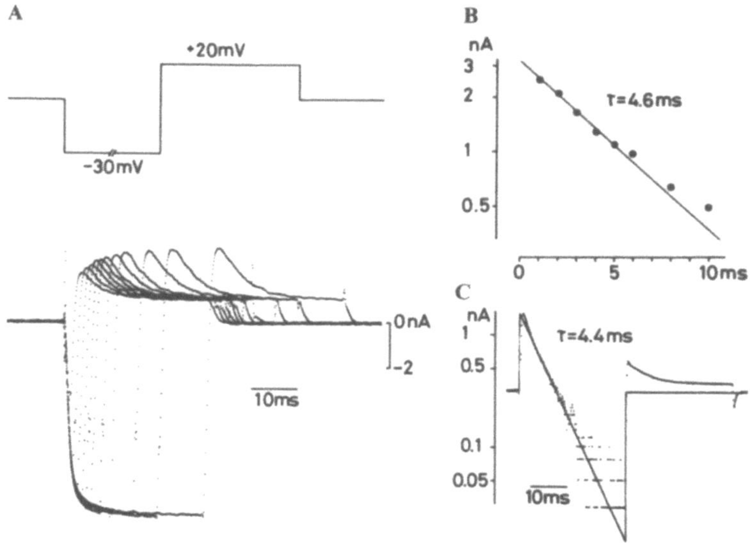

Fig. 3. **A** *Upper panel* indicates the potential protocol in which the hyperpolarising prepulse was progressively prolonged, and the *lower panel* the superimposed current traces corresponding, from left to right, to hyperpolarising pulses of duration 1, 2, 3, 4, 5, 6, 8, 10, 15, 20 and 30 ms, respectively. Internal $[K^+]=150$ m*M*; internal free $[Mg^{2+}]=15$ μ*M*; external $[K^+]=40$ m*M*. Figure in the pulse protocol indicates a voltage deviation from the holding potential, which was set to the zero current potential. Capacitive currents were removed by subtracting current traces recorded in K^+-free external solution during the same voltage step. **B** Instantaneous outward current was obtained by extrapolating the best fitted curve to the onset of the depolarising pulse and subtracted from its maximum value. The resulting value was plotted against the duration of the preceding pulse on a semi-logarithmic plot. A line giving a time constant of 4.6 ms was fitted by eye. **C** Semi-logarithmic plot of the inward current during the 30-ms hyperpolarising pulse. The increase of the inward current can be fitted with a single exponential function having a time constant of 4.4 ms. (From Saigusa and Matsuda 1988, with permission)

linear I–V relation obtained in the whole-cell current experiments. In these experiments, $Tris^+$ ion, which was not supposed to affect the inwardly rectifying K^+ channel, was used as a substitute for Na^+. As a matter of fact, $Tris^+$ ion decreases the amplitude of inward unitary currents without affecting the kinetics or outward unitary currents, producing outward rectification in single-channel conductance with 30 m*M* K^+, 120 m*M* $Tris^+$ pipette solution (H. Matsuda, unpublished observation). Why a linear I–V relationship was obtained in the whole-cell current recordings instead of an outwardly rectifying I–V relation is not clear. Intracellular Mg^{2+}, whose concentration was supposed to be higher than 15 μ*M*, may have played some role. This is suggested by the fact that 6 of 21 cells showed inward rectification. The linear instantaneous I–V relation obtained in the whole-cell current experiments, including some ambiguity in retrospect, led us to record outward unitary currents (Matsuda et al. 1987).

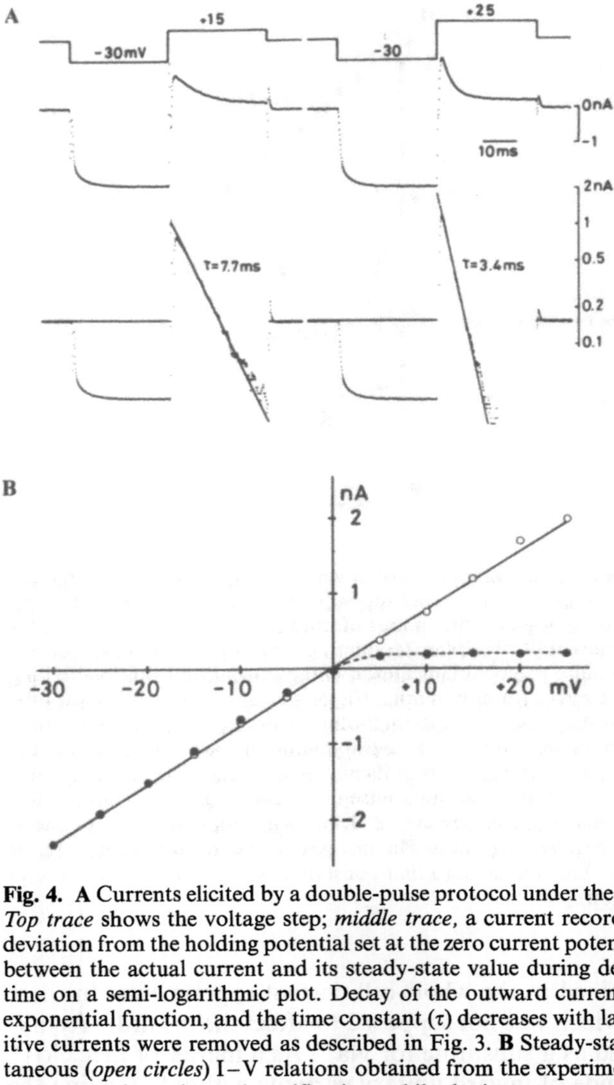

Fig. 4. A Currents elicited by a double-pulse protocol under the same conditions as in Fig. 3. *Top trace* shows the voltage step; *middle trace,* a current record. Figure indicates a voltage deviation from the holding potential set at the zero current potential. *Bottom trace,* difference between the actual current and its steady-state value during depolarisation plotted against time on a semi-logarithmic plot. Decay of the outward current can be fitted with a single exponential function, and the time constant (τ) decreases with larger depolarisations. Capacitive currents were removed as described in Fig. 3. **B** Steady-state (*filled circles*) and instantaneous (*open circles*) I–V relations obtained from the experiment shown in **A**. Steady-state I–V curve shows inward rectification, while the instantaneous I–V curve is almost linear, with a slope conductance of 80 nS. (From Matsuda et al. 1987)

Ohmic Current–Voltage Relations for the Single Channel in the Absence of Internal Mg^{2+}

The inwardly rectifying K^+ channel responsible for the resting conductance was the first cardiac K^+ channel that was examined in single-channel recordings (Trube et al. 1981). Gating kinetics during hyperpolarisation was studied earlier in detail (Kameyama et al. 1983; Sakmann and Trube 1984b); inward currents show characteristically long-lasting openings (lower panel of Fig. 6A). As reported previously (Sakmann and Trube 1984a), outward single-channel currents through the inwardly rectifying K^+ channel were not recorded in the cell-attached configuration. Transient outward currents, however, appeared in response to voltage steps to levels more positive than E_K, 2–5 min after the cell membrane was ruptured by crushing the tip of another patch pipette against the cell on the bottom of the recording chamber, and the inner surface of the patch came into contact with Ca^{2+}-free, Mg^{2+}-free, high-K^+ solution (Fig. 5; open cell-attached configuration).

Such currents are illustrated in the upper panels of Fig. 6A, together with inward currents recorded in a steady-state condition (lower panels), at pipette K^+ concentrations of 150 m*M* (left) and 40 m*M* (right). Several findings suggest that these outward unitary currents flow through the inwardly rectifying K^+ channel. (1) Single outward currents were not recorded from patches without inwardly rectifying K^+ channels. Overlapped outward currents were recorded from patches showing multichannel currents in the inward direction and the maximum number of overlaps coincided, in most cases. (2) Whether

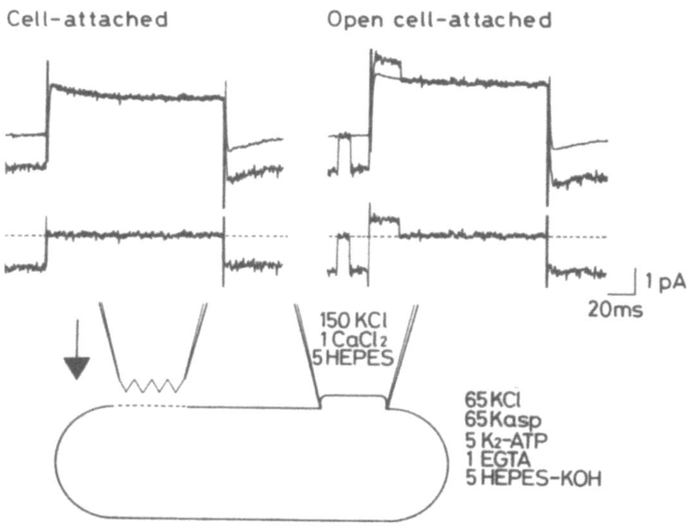

Fig. 5. Single-channel currents recorded from an inwardly rectifying K^+ channel in the cell-attached (*left*) and open cell-attached configuration (*right*) at a pipette K^+ concentration of 150 m*M*. Patch membrane was held at −68 mV and depolarised to +30 mV. *Thin trace in the upper panel* is the average of current traces without events. *Middle panel* shows current traces after subtraction of the average from traces with events. *Dotted line* indicates zero current level. As shown in the *lower panel*, the cell was perfused with high-K^+, divalent ion-free solution. In this solution, the cell membrane was ruptured by crushing the tip of another patch pipette against the cell on the glass bottom of the recording chamber. Composition of pipette and perfusing solutions is shown in m*M* (*Kasp*, K aspartate). All records in the following figures were obtained in the open cell-attached configuration at pipette and internal K^+ concentrations of 150 m*M*. Currents were filtered using a Bessel filter with a −3 dB corner frequency of 0.8–1 kHz and sampled every 0.2 ms, unless otherwise indicated

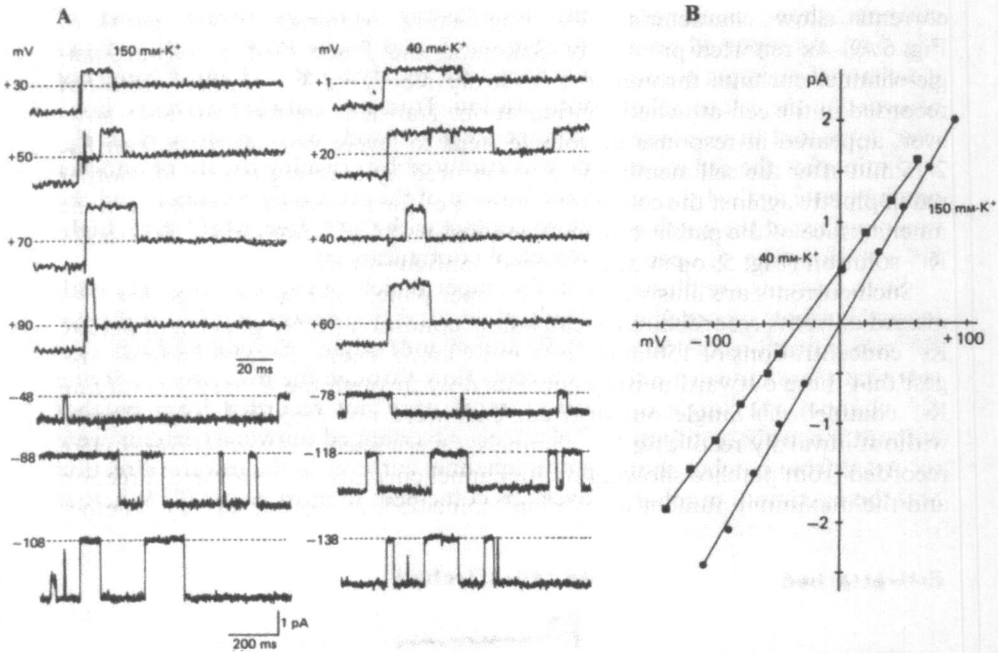

Fig. 6. A Single-channel currents recorded from an inwardly rectifying K^+ channel in the open cell-attached configuration at pipette K^+ concentrations of 150 m*M* (*left*) and 40 m*M* (*right*). Numbers to the *left* of each current trace refer to (*upper four panels*) the potential levels during the depolarising steps from −48 mV (150 m*M* K^+) and −78 mV (40 m*M* K^+) or (*lower three panels*) holding potential. *Dotted line* indicates zero current level. Sample rate, 5 kHz (*upper panels*) and 1 kHz (*lower panels*). **B** Single-channel I−V relationships obtained from the same patches as in **A.** Straight lines were drawn by eye. (From Matsuda 1988)

the record obtained during depolarisation began with an open or closed state depended usually on the state immediately before steps to positive potentials. When a depolarising pulse was applied during the open state, the channel stayed open at the onset of the pulse and closed during the pulse in most cases. When the pulse was applied during the closed state, either the outward current was not recorded or appeared with a delay (+50 mV at 150 m*M* K^+ and +40 mV at 40 m*M* K^+ in Fig. 6 A). (3) Internal tetraethylammonium (TEA^+) ions blocked both outward and inward currents (Fig. 7).

The I−V relation shows both inward and outward currents, which were well fitted with a line which crosses the voltage axis at around E_K (Matsuda et al. 1987). Using voltage ramps, Vandenberg (1987) also obtained a linear single-channel I−V relation from inside-out patches exposed to Mg^{2+}-free solution. Ohmic I−V relation was obtained with 40 m*M* K^+ pipette solution as well as with 150 m*M* K^+ solution (Fig. 6 B). It should be noted that the unitary conductance for the outward current was altered by changing the external K^+ concentration at a constant internal K^+ level. The two lines should cross each other if outward currents are measured at more positive potentials.

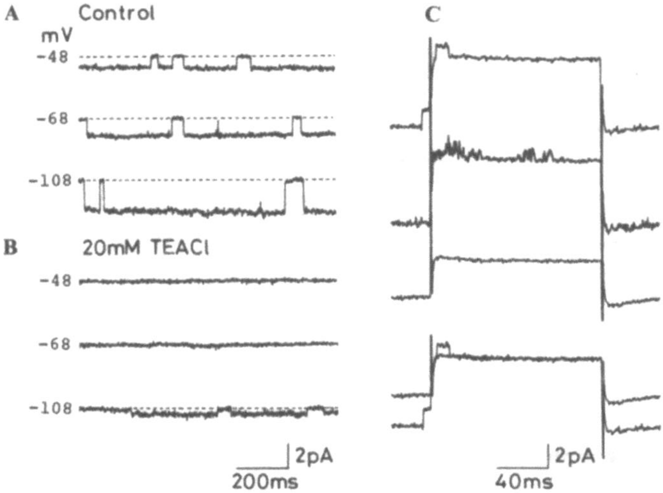

Fig. 7A–C. Effects of internal tetraethylammonium chloride (TEA Cl) on single-channel currents through the inwardly rectifying K^+ channel in the open cell-attached configuration. **A** and **B** Steady-state inward currents during the control (**A**) and after (**B**) adding 20 m*M* TEA Cl to high-K^+, divalent cation-free solution. **C** Current traces in response to a voltage step from −68 to +50 mV obtained (from top) before, 40 s and 90 s after addition of 20 m*M* TEA Cl. Note that both inward and outward currents became very rapid (*second trace*) and then suppressed (*third trace*). Control and third trace were superimposed in the *bottom trace*. Capacitive and leakage currents were not subtracted. Sampling frequency, the same as in Fig. 6

Voltage-Dependent Channel Kinetics Independent of Internal Mg^{2+}

As shown in the upper panel of Fig. 6A, the channel usually closes during the depolarising pulse (130 ms in duration) in the absence of internal Mg^{2+}. As a result, average currents decay during depolarisation, and a time course of current decay is fitted with a single-exponential function (Fig. 8). This indicates that the channel closes during depolarisation through a voltage-dependent gating mechanism operating independently of the Mg^{2+} block, resulting in inward rectification of the steady state. Gating kinetics, independent of internal Mg^{2+}, was also noted in the muscarinic receptor-operated K^+ channel (Horie and Irisawa 1987).

The open times of the channel in the outward direction tended to be prolonged with time after opening the cell to bath solution, causing a slow increase in the time constant. In the case shown in Fig. 8, the time constant of the exponential decay of the average current (bottom trace) increased from 7.8 ms (left column) to 75.3 ms (right column) during perfusion for 40 min. This result suggests that dialysis of the open cell washed away a soluble gating component related to channel closing. Nevertheless, the time constant appeared to be

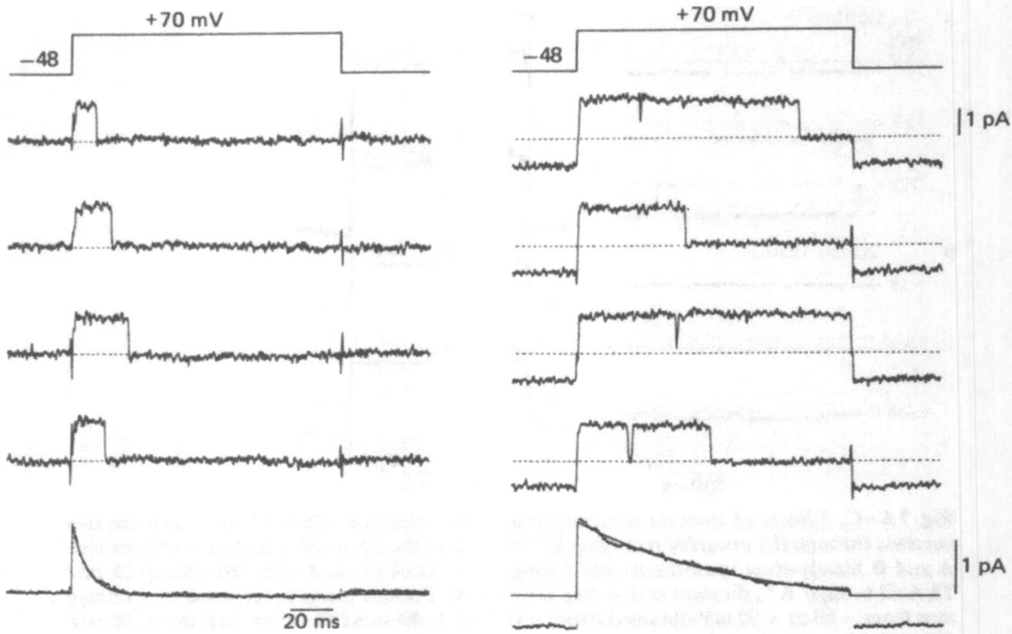

Fig. 8. In each column, the *top trace* shows the voltage step, the *middle traces* sample records, and the *bottom trace* the average current. Records in the *left column* were obtained within 15 min after opening the cell membrane and records in the *right column* from the same patch 40 min after the first recording. During the first recording, current traces without events were so few that current traces with inward events but no outward events were averaged for subtraction. Thus, the holding current was removed by the subtraction procedure. *Dotted line* indicates zero current level. Average current was obtained from 113 (*left*) and 100 (*right*) frames and is plotted by points. The time constant of the exponential decay (shown by the *continuous line*) was 7.8 ms (*left*) and 75.3 ms (*right*). (From Matsuda 1988)

voltage dependent (Table 1 of Matsuda 1988), and the steady-state level of the average current was negligibly small (Fig. 8). It is suggested that the closing rate constant increases as the membrane potential becomes more positive and that the rate constant for channel opening is much smaller compared with the rate constant for channel closing.

Voltage-Dependent Blockage by Internal Mg^{2+}

Mg^{2+} Acts as an Open Channel Blocker

The results presented so far have described the properties of outward unitary currents through the inwardly rectifying K^+ channel when the patch was exposed to artificial divalent-free internal solution: the channel conductance is ohmic, and the channel closes through intrinsic gating kinetics to cause steady-state inward rectification when the membrane potential is made positive to E_K. The failure to observe outward currents in the cell-attached configuration (Fig. 5) suggests an additional mechanism responsible for the rectification of this channel, the voltage-dependent block of the channel by intracellular substances. One such substance is Mg^{2+}.

Figure 9 shows the outward current during a clamp pulse stepped to +70 mV from a holding potential of −48 mV. In the absence of internal Mg^{2+}, full-size single channel currents were observed on depolarisation, usually followed by a closed state during the 130-ms pulse. In the presence of Mg^{2+} on the cytoplasmic side, the channel no longer stayed fully open or fully closed. Instead, the outward open channel showed two discrete, equally spaced, subconductance levels and fluctuated between four levels including these sublevels. As the Mg^{2+} concentration was increased, the channel stayed at the lower levels more often, and the fluctuations became faster. The noisy trace at 100 μ*M* Mg^{2+} is ascribed to failure of the recording system to follow the fast transitions. Note that internal Mg^{2+} prolonged the channel opening itself so that channel activity was seen throughout the pulse. This effect is completely reversible: the channel can close after washing out the internal Mg^{2+}.

As may be anticipated from experiments with other blocking ions, blockage by internal Mg^{2+} increases both with depolarisation of the membrane and with the concentration of Mg^{2+}. Figure 10 A shows the mean open-channel I−V relations at different Mg^{2+} concentrations. The chord conductance was decreased at potentials positive to E_K, and a negative slope of the I−V relation was observed with 2 and 10 μ*M* Mg^{2+}. The mean open-channel current was normalised to that measured in the absence of Mg^{2+}. The fractional current-Mg^{2+} concentration curve was fitted by a one-to-one binding curve (Fig. 10 B). Dissociation constants averaged in 3−4 experiments at 25 °C are 10.5 ± 1.7 μ*M* at +30 mV, 3.2 ± 0.3 μ*M* at +50 mV, 1.6 ± 0.2 μ*M* at +70 mV and 0.7 ± 0.1 μ*M* at +90 mV. The voltage dependence of dissociation constants gave a value for the fractional electrical distance of the Mg^{2+} binding site of 0.59 (Woodbury 1971). This value is close to 0.63 in the muscarinic receptor-operated K^+ channel (Horie and Irisawa 1989). A value of 0.32 was reported for the ATP-regulated K^+ channel (Horie et al. 1987).

The cytosolic free Mg^{2+} concentration in cardiac muscle is reported to range from 0.5 to 3.5 m*M* (Hess et al. 1982; Gupta et al. 1984; Blatter and McGuigan 1986). This concentration is sufficient to depress outward currents (Matsuda et al. 1987; Vandenberg 1987) and explains why outward currents were not recorded in the cell-attached configuration. Internal Mg^{2+}, examined up to 1 m*M*, had no effect on the steady-state inward currents.

The outward currents appeared within 2−5 min of opening the cell membrane. At this time, in Mg^{2+}-free solution, the outward current blocked partially as shown in Fig. 9 was never observed. The outward current in the fully open state appeared from the very first. This can be explained by assuming that there might be an intracellular blocker (or blockers) other than Mg^{2+}, which takes more time to be washed out than Mg^{2+} does. Alternatively, a soluble

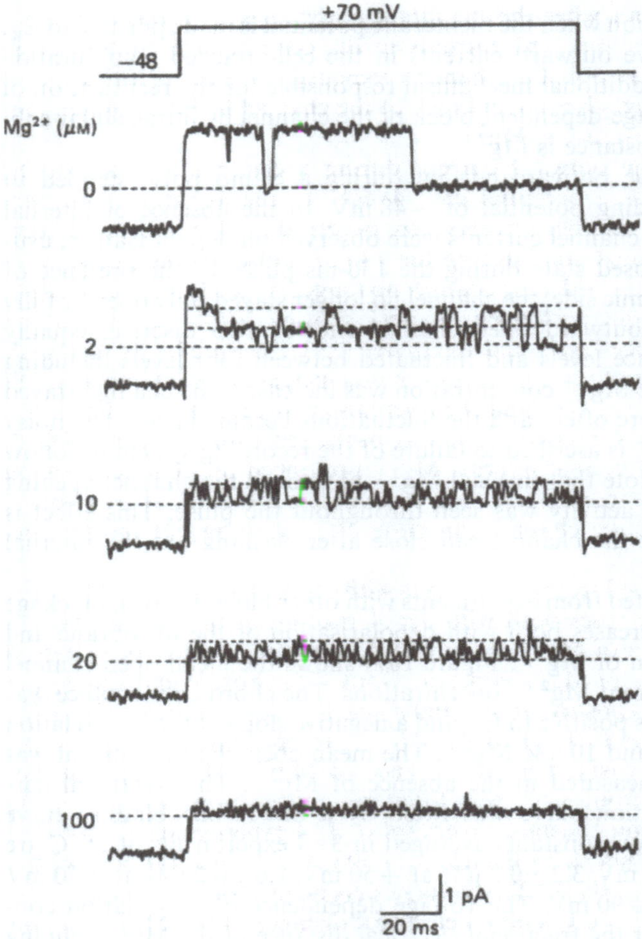

Fig. 9. Effect of internal Mg^{2+} on the outward single-channel current. *Top trace* shows the voltage step. After recording the control trace while perfusing the open-cell patch membrane with high-K^+, divalent cation-free solution, the Mg^{2+} concentration of the perfusing solution was increased progressively. Then Mg^{2+} was washed out, and recording at 0 and 2 μ*M* Mg^{2+} was repeated. Traces for illustration were obtained 55 min (0 μ*M*), 65 min (2 μ*M*), 30 min (10 μ*M*), 40 min (20 μ*M*) and 50 min (100 μ*M*) after opening the cell membrane. *Longer dashed lines* indicate zero current level, and the *shorter ones* show sublevels of the open-channel current. Note that Mg^{2+} had no effect on the inward current at the holding potential. Mg^{2+} concentration was buffered using EDTA. (From Matsuda 1988)

gating component could be washed away to some degree for the outward current to be prolonged and detected.

Disappearance of inward rectification by exposing the internal surface of the membrane vesicle to a Mg^{2+}-free solution and restoration by adding Mg^{2+} to the internal bathing solution at a free ion concentration higher than

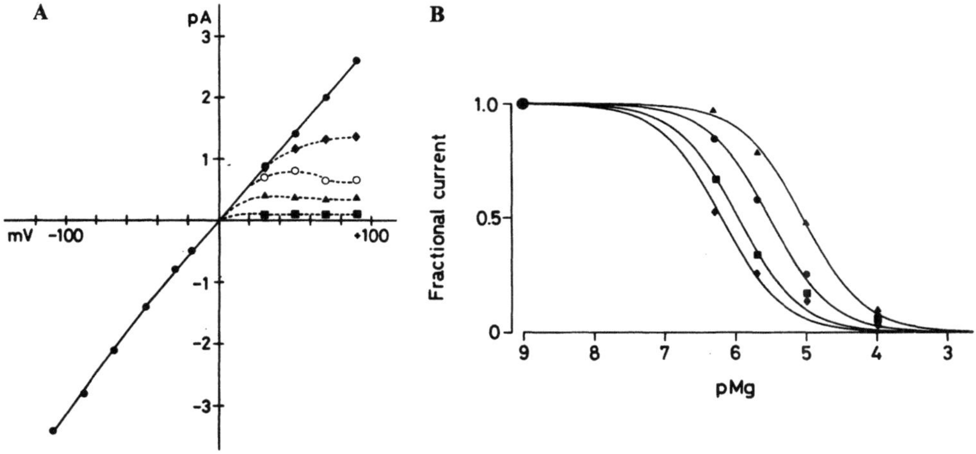

Fig. 10. **A** Mean open-channel current–voltage relationships at 0 (*closed circles*), 0.5 (*diamonds*), 2 (*open circles*), 10 (*triangles*) and 100 (*sqaures*) μM internal Mg^{2+}. Outward mean open-channel currents were calculated from 20 to 50 frames. Unit amplitude of inward currents in the presence of 100 μM Mg^{2+} were superposable to those in the absence of Mg^{2+} and omitted in this figure. See Fig. 14 for data at 100 μM Mg^{2+}. In this patch, a slope conductance was increased slightly at more negative potentials than −50 mV. **B** Fractional current-Mg^{2+} concentration curves obtained from the data shown in **A**: at +30 (*triangles*), +50 (*circles*), +70 (*squares*) and +90 (*diamonds*) mV. They were fitted by a one-to-one binding curve with a dissociation constant of (from left) 0.6, 1.4, 3.0 and 8.8 μM, respectively. Data were obtained at 25 °C

14 μM have been reported in the inwardly rectifying K^+ channel from rat skeletal muscle (Burton and Hutter 1989).

Substate Conductance Revealed by Internal Mg^{2+}

The most characteristic finding in the Mg^{2+} block of the inwardly rectifying K^+ channel is the substate behaviour seen easily with low internal Mg^{2+}. Figure 11 A shows the outward current records at different voltages in the presence of 2 μM Mg^{2+}. The channel fluctuated between well-defined sublevels at potentials more positive than +50 mV. As the voltage is made increasingly positive, the probability of observing the one-third level increases progressively, while that of observing the two-thirds or fully open level decreases. Sublevels with one-third or two-thirds of the unitary amplitude were observed spontaneously in the inward current (Fig. 11 B). They usually lasted for less than 50 ms and occurred much less frequently in comparison with the fully open level. Internal Mg^{2+} never increased the frequency of such sublevels in the inward direction, in contrast to the outward direction. The I–V relation for each level obtained from the patches shown in Fig. 11 A and B is linear in the voltage range examined (between −108 and +90 mV), and all fitted lines

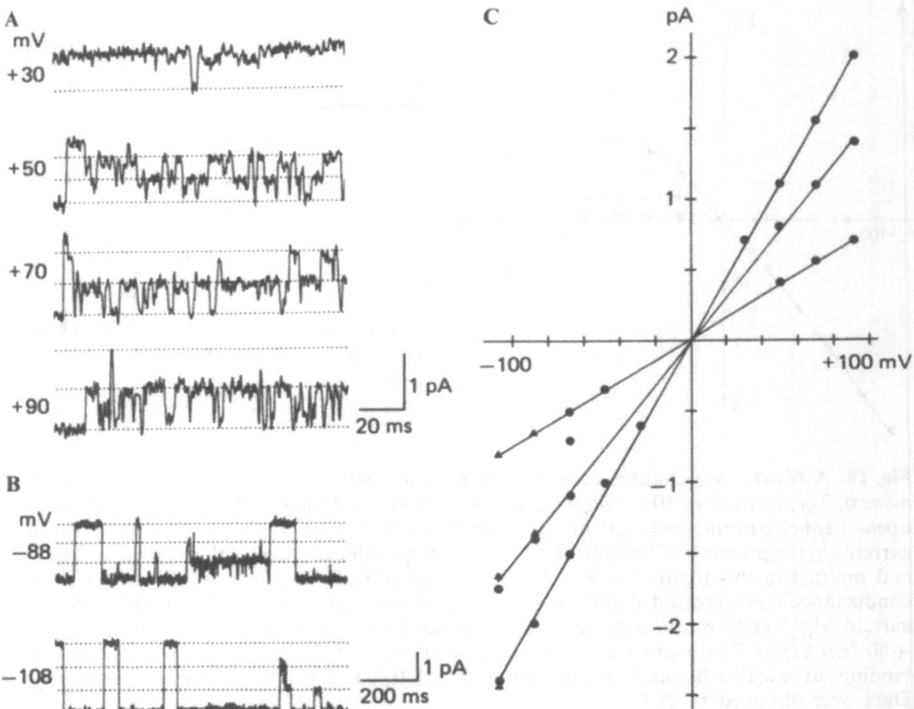

Fig. 11. **A** Outward currents at diffeent voltages. *Dotted lines* indicate, from the bottom, zero current level and sublevels with one-third and two-thirds of the unit amplitude (except at +30 mV). Mg^{2+} concentration on the cytoplasmic side = 2 μ*M*. **B** Sublevels observed in the inward direction in the absence of internal Mg^{2+}. *Dotted lines* indicate, from the top, zero current level and sublevels with one-third and two-thirds of the unit amplitude. **C** Current–voltage relation of open-channel substates. *Circles,* data from the patch shown in **A**; *diamonds,* data from the patch shown in **B**. For the patch in **A**, data at positive potentials were obtained in the presence of 2 μ*M* Mg^{2+} and those at negative potentials were collected at Mg^{2+} concentrations of 0 and 100 μ*M*. Lines were fitted by eye. (From Matsuda 1988)

crossed the voltage axis near 0 mV, the E_K predicted from the transmembrane K^+ concentration (Fig. 11 C). The slope conductances of the intermediate states are one-third and two-thirds of the fully open channel conductance.

The appearance, when the channel is open, of four equally spaced conductance levels (including the zero current level) suggests that the cardiac inwardly rectifying K^+ channel consists of three identical conducting units. They usually function cooperatively to form a single channel, but Mg^{2+} enters and plugs up each subunit to produce the substate behaviour at positive potentials. If the three subunits are blocked independently by internal Mg^{2+}, the open-state occupancies of each current level should conform to a binomial distribution. Fractions of the total time for which the current stayed at each level were measured at various voltages and Mg^{2+} concentrations and plotted against

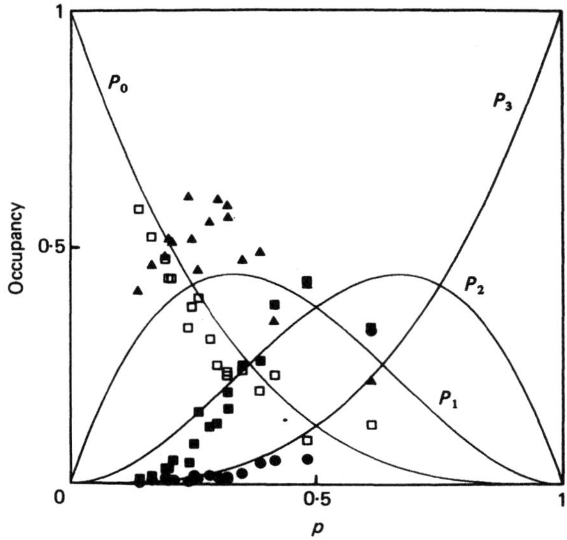

Fig. 12. Distribution of sublevels obtained in five experiments. Occupancies at various Mg^{2+} concentrations and voltages were plotted against the observed open probability of subunits (p). *Open squares,* zero current level; *triangles,* one-third level; *filled squares,* two-thirds level; *circles,* fully open level. Theoretical curves were drawn from the following equations: $P_0=(1-p)^3$, $P_1=3p(1-p)^2$, $P_2=3p^2(1-p)$ and $P_3=p^3$. (From Matsuda 1988)

the open probability of subunits together with theoretical curves (Fig. 12). P_0, P_1, P_2 and P_3 represent the occupancy at current levels of zero, one-third, two-thirds and fully open, respectively. Data points for the one-third level tend to be located above the theoretical curve. However, the deviation from the theoretical curves seems, on the whole, not so large. Sublevels with one-third or two-thirds of the unit amplitude were induced in the inward current by the external application of Cs^+ or Rb^+, though not in all channels, which was different from the case in the Mg^{2+} block (Matsuda et al. 1989). The distribution of the current levels showed a reasonable agreement with the binomial theorem as in the Mg^{2+} block.

The kinetic properties of blockage accompanied by the substate behaviour were studied based on the binomial scheme (Matsuda 1988; Matsuda et al. 1989). If the block of each subunit is described as:

$$\mathrm{O} \underset{\lambda}{\overset{\mu}{\rightleftharpoons}} \mathrm{B}$$

where O and B are the open and blocked states of each subunit, λ is the first-order unblocking rate and μ the second-order blocking rate, the open-state probability of subunits p is expressed as $\lambda/(\lambda+\mu)$, and transitions between substrates during the open state of the channel can be described as:

$$\mathrm{O}_3 \underset{\lambda}{\overset{3\mu}{\rightleftharpoons}} \mathrm{O}_2 \underset{2\lambda}{\overset{2\mu}{\rightleftharpoons}} \mathrm{O}_1 \underset{3\lambda}{\overset{\mu}{\rightleftharpoons}} \mathrm{O}_0$$

where O_0, O_1, O_2 and O_3 are the substates in which all, two, one and none of three subunits are blocked, respectively. In this scheme, the mean dwelling times in the substates are given:

$$1/\tau_0 = 3\lambda$$

$$1/\tau_1 = 2\lambda + \mu$$

$$1/\tau_2 = \lambda + 2\mu$$

$$1/\tau_3 = 3\mu$$

where τ_0, τ_1, τ_2 and τ_3 represent the mean lifetimes in O_0, O_1, O_2 and O_3, respectively.

The rate constants are overdetermined in terms of the binomial model; five measurable quantities, τ_0, τ_1, τ_2, τ_3 and p yield calculations of only two rate constants. Therefore the validity of this model can be tested by calculating the rate constants from some of the measurements and comparing the values which were not used for calculation with the values predicted from the rate constants. The dwell-time histogram in each substrate could be fitted by a single-exponential function (not shown; see Fig. 8 of Matsuda 1988). In Mg^{2+} block, λ and μ which give the best fit to τ_1, τ_2 and p were determined by a least-squares method. The τ_3 predicted from the rate constants was not so different from the observed value, but the predicted τ_0 was larger by 2–3 times than the observed value (Table 2 of Matsuda 1988). Thus, the kinetics of Mg^{2+} blockage seems not so simple as described above. There may be some interaction between subunits. A better agreement was found in Cs^+ and Rb^+ block than in Mg^{2+} block.

Nevertheless, the rate constants can explain the relaxation of outward currents during depolarisation in the presence of internal Mg^{2+}. As shown in Fig. 9, the fully open current was observed frequently at the beginning of positive pulses, especially at lower Mg^{2+} concentrations, and then it declined to lower substate levels. The average currents from such traces decline with time. Because in the presence of internal Mg^{2+} the channel shows rapid transitions between substate levels throughout the depolarising clamp pulse of 130 ms, the relaxation of the average current is ascribed to the voltage-dependent Mg^{2+} block rather than to the intrinsic gating mechanisms. Figure 13 shows the average currents obtained at different voltages in the presence of 2 μ*M* Mg^{2+}. Neglecting a very slow decline of the current, the time course of relaxation was fitted with a single-exponential function. The time constant (τ) decreased with larger depolarisation and increasing Mg^{2+} concentration. The observed time constants are in good agreement with the time constants for block (τ_B) predicted from the relation, $\tau_B = 1/(\lambda + \mu)$, favouring the binomial model.

The average outward current with 2 μ*M* Mg^{2+} reached a steady level of sizeable amplitude after relaxation. This is in contrast to the average current without Mg^{2+}, which eventually approached the zero current level (compare Fig. 8 with Fig. 13). Such a difference results from the prolongation of the channel opening by Mg^{2+} as noted before. The simplest explanation of these

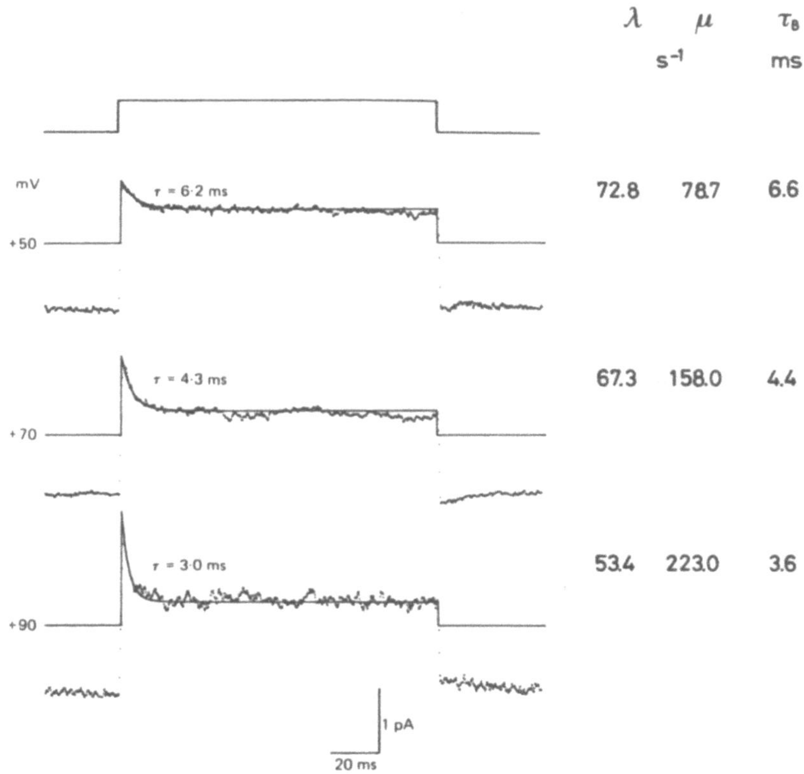

Fig. 13. Relaxation of average current induced by voltage-dependent Mg^{2+} block. *Top trace* shows the voltage step from a holding potential of −48 mV to voltages as indicated. Average from 74 frames at +50 mV, 49 frames at +70 mV and 25 frames at +90 mV. *Continuous line* during hyperpolarisation indicates zero current level. Note a sizeable steady-state outward current. Numbers to the *right* of each trace indicate unblocking (λ) and blocking (μ) rates obtained as described in the text and the time constant of current decay predicted from these rate constants. (From Matsuda 1988)

findings is to assume that the channel is stuck in the open state by the presence of Mg^{2+} in at least one subunit of the single channel. Thus, a partial block of the channel by low Mg^{2+} produces a larger steady-state current than without Mg^{2+}. It is unlikely that this mechanism works at the intracellular free Mg^{2+} concentrations reported so far (0.5–3.5 m*M*).

Although the effect of Mg^{2+} varied to some degree among the individual experiments, there seemed to be no dependence of λ on Mg^{2+}, while μ increased linearly with increasing Mg^{2+} (Table 2 of Matsuda 1988). The blocking rate increased with more positive potentials, while the unblocking rate seemed to decrease slightly. This result corresponds well with the idea of the open channel blocker. A blocking rate constant of 5×10^7 s^{-1} M^{-1} (at +70 mV) is near to 1.7×10^7 s^{-1} M^{-1} in the Mg^{2+} blockage of the ATP-regulated K^+ channel (Horie et al. 1987) and to $1.4 \times 10^7\ s^{-1}\ M^{-1}$ in the Mg^{2+} blockage of the cation channel in the ganglion cell (Nowak et al. 1984). Voltage dependence of the blocking rate is as expected for an inner energy barrier located 30% across the electrical field (Matsuda 1988).

Comparison with the Mg^{2+} Block in Other Cardiac K^+ Channels

Unidirectional and voltage-dependent blockage by internal Mg^{2+} was first noted in the ATP-regulated K^+ channel by Horie et al. (1987), who analysed outward open-channel current using noise analysis. The open-channel blockage by Mg^{2+} was observed also in the muscarinic receptor-operated K^+ channel (Horie and Irisawa 1987, 1989) and the Na^+-activated K^+ channel (M. Horie, personal communication) and seems to be a common mechanism causing inward rectification of cardiac K^+ channels.

In Fig. 14, the mean open-channel I–V relation with 100 μM Mg^{2+} was presented (squares; the same data as in Fig. 10A), together with that in the ATP-regulated K^+ channel with 500 μM Mg^{2+} (circles; reproduced from Horie et al. 1987) and that in the muscarinic receptor-operated K^+ channel with 100 μM Mg^{2+} (triangles; reproduced from Horie and Irisawa 1989). As is evident in this figure, the inwardly rectifying K^+ channel is the most sensitive

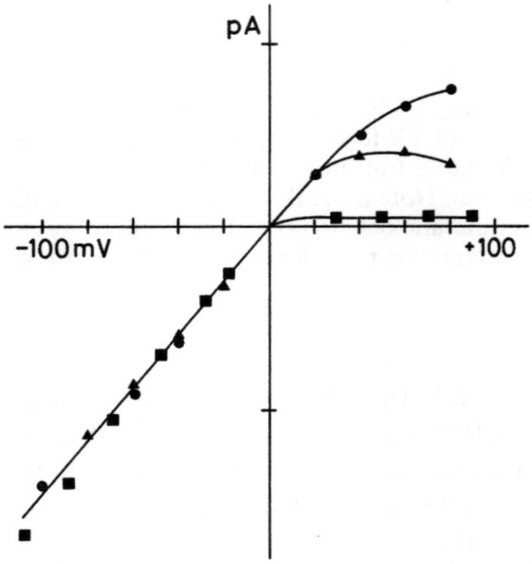

Fig. 14. Mean open-channel current–voltage relationships in the ATP-regulated K^+ channel (*circles*, 500 μM Mg^{2+}, 34 °C), the muscarinic receptor-operated K^+ channel (*triangles*, 100 μM Mg^{2+}, 22–24 °C, guinea pig atrial cells) and the inwardly rectifying K^+ channel (*squares*, 100 μM Mg^{2+}, 25 °C). Scales on current axis correspond to ±5.22 pA for the ATP-regulated K^+ channel, ±2.8 pA for the muscarinic receptor-operated K^+ channel and ±2.0 pA for the inwardly rectifying K^+ channel. (Data for the ATP-regulated K^+ channel taken from Horie et al. 1987 and for the muscarinic receptor-operated K^+ channel from Horie and Irisawa 1989, with permission)

to Mg^{2+} among these cardiac K^+ channels. The dissociation constant at +40 mV was 2.0 m*M* in the ATP-regulated K^+ channel (Horie et al. 1987) and 0.29 m*M* in the muscarinic receptor-operated K^+ channel (Horie and Irisawa 1989).

Sublevels equivalent to one-third and two-thirds of the unit amplitude were also noted in the muscarinic receptor-operated K^+ channel for both the outward and inward currents (Horie and Irisawa 1989).

Conclusion

Intracellular Mg^{2+} acts as an open channel blocker of cardiac K^+ channels and produces inward rectification at physiological concentrations. The Mg^{2+} blockage is not the only mechanism responsible for inward rectification. The inwardly rectifying K^+ channel and the muscarinic receptor-operated K^+ channel are also gated by voltage, so that even in the absence of internal Mg^{2+} the open state probability is reduced when the membrane potential is made positive to E_K. Internal blocking ions or substances other than Mg^{2+} are also suggested. Substate behaviour revealed by low internal Mg^{2+} indicates that the channel is composed of three identical conducting units and that each unit is blocked by Mg^{2+}. Blockage of each subunit may not be completely independent.

Acknowledgements. I thank Professor H. Irisawa for comments on the manuscript and Dr M. Horie and the Physiological Society for permission to reproduce illustrations.

References

Armstrong CM (1975) Potassium pores of nerve and muscle membranes. In: Eisenman G (ed) Membranes. Dekker, New York, pp 325–358 (A series of advances, vol 3)

Blatter LA, McGuigan JAS (1986) Free intracellular magnesium concentration in ferret ventricular muscle measured with ion-selective micro-electrodes. Q J Exp Physiol 71:467–473

Burton FL, Hutter OF (1989) Properties of 'inwardly rectifying' potassium channels from rat muscle in absence of 'intracellular' Mg^{2+}. J Physiol (Lond) 409:51P

Callewaert G, Vereecke J, Carmeliet E (1986) Existence of a calcium-dependent potassium channel in the membrane of cow cardiac Purkinje cells. Pflügers Arch 406:424–426

Gupta RK, Gupta P, Moore RD (1984) NMR studies of intracellular metal ions in intact cells and tissues. Annu Rev Biophys Bioeng 13:221–246

Hagiwara S (1983) Anomalous rectification. In: Membrane potential-dependent ion channels in cell membrane. Phylogenetic and developmental approaches. Raven, New York, pp 65–79

Hamill OP, Marty A, Neher E, Sakmann B, Sigworth FJ (1981) Improved patch-clamp techniques for high-resolution current recording from cells and cell-free membrane patches. Pflügers Arch 391:85–100

Hess P, Metzger P, Weingart R (1982) Free magnesium in sheep, ferret and frog striated muscle at rest measured with ion-selective micro-electrodes. J Physiol (Lond) 333:173–188

Hille B, Schwarz W (1978) Potassium channels as multi-ion single-file pores. J Gen Physiol 72:409–442
Hodgkin AL, Horowicz P (1959) The influence of potassium and chloride ions on the membrane potential of single muscle fibres. J Physiol (Lond) 148:127–160
Horie M, Irisawa H (1987) Rectification of muscarinic K^+ current by magnesium ion in guinea pig atrial cells. Am J Physiol 253:H210–H214
Horie M, Irisawa H (1989) Dual effects of intracellular magnesium on muscarinic potassium channel current in single guinea-pig atrial cells. J Physiol (Lond) 408:313–332
Horie M, Irisawa H, Noma A (1987) Voltage-dependent magnesium block of adenosine-triphosphate-sensitive potassium channel in guinea-pig ventricular cells. J Physiol (Lond) 387:251–272
Isenberg G, Klöckner U (1982) Isolated bovine ventricular myocytes. Characterization of the action potential. Pflügers Arch 395:19–29
Kameyama M, Kiyosue T, Soejima M (1983) Single channel analysis of the inward rectifier K current in the rabbit ventricular cells. Jpn J Physiol 33:1039–1056
Kameyama M, Kakei M, Sato R, Shibasaki T, Matsuda H, Irisawa H (1984) Intracellular Na^+ activates a K^+ channel in mammalian cardiac cells. Nature 309:354–356
Katz B (1949) Les constantes électriques de la membrane du muscle. Arch Sci Physiol 3:285–300
Matsuda H (1988) Open-state substructure of inwardly rectifying potassium channels revealed by magnesium block in guinea-pig heart cells. J Physiol (Lond) 397:237–258
Matsuda H, Noma A (1984) Isolation of calcium current and its sensitivity to monovalent cations in dialysed ventricular cells of guinea-pig. J Physiol (Lond) 357:553–573
Matsuda H, Saigusa A, Irisawa H (1987) Ohmic conductance through the inwardly rectifying K channel and blocking by internal Mg^{2+}. Nature 325:156–159
Matsuda H, Matsuura H, Noma A (1989) Triple-barrel structure of inwardly rectifying K^+ channels revealed by Cs^+ and Rb^+ block in guinea-pig heart cells. J Physiol (Lond) 413:139–157
Noma A (1983) ATP-regulated K^+ channels in cardiac muscle. Nature 305:147–148
Nowak L, Bregestovski P, Ascher P, Herbet A, Prochiantz A (1984) Magnesium gates glutamate-activated channels in mouse central neurones. Nature 307:462–465
Saigusa A, Matsuda H (1988) Outward currents through the inwardly rectifying potassium channel of guinea-pig ventricular cells. Jpn J Physiol 38:77–91
Sakmann B, Trube G (1984a) Conductance properties of single inwardly rectifying potassium channels in ventricular cells from guinea-pig heart. J Physiol (Lond) 347:641–657
Sakmann B, Trube G (1984b) Voltage-dependent inactivation of inward-rectifying single-channel currents in the guinea-pig heart cell membrane. J Physiol (Lond) 347:659–683
Sakmann B, Noma A, Trautwein W (1983) Acetylcholine activation of single muscarinic K^+ channels in isolated pacemaker cells of the mammalian heart. Nature 303:250–253
Shibasaki T (1987) Conductance and kinetics of delayed rectifier potassium channels in nodal cells of the rabbit heart. J Physiol (Lond) 387:227–250
Standen NB, Stanfield PR (1978) Inward rectification in skeletal muscle: a blocking particle model. Pflügers Arch 378:173–176
Stanfield PR (1988) Intracellular Mg^{2+} may act as a co-factor in ion channel function. Trends Neurosci 11:475–477
Trube G, Sakmann B, Trautwein W (1981) Inward rectifying potassium currents recorded from isolated heart cells by the patch clamp method. Pflügers Arch 391:R28
Vandenberg CA (1987) Inward rectification of a potassium channel in cardiac ventricular cells depends on internal magnesium ions. Proc Natl Acad Sci USA 84:2560–2564
Woodbury JW (1971) Eyring rate theory model of the current–voltage relationships of ion channels in excitable membranes. In: Hirschfelder JO (ed) Chemical dynamics. Papers in honor of Henry Eyring. Wiley, New York, pp 601–617
Yue DT, Marban E (1988) A novel cardiac potassium channel that is active and conductive at depolarized potentials. Pflügers Arch 413:127–133

Regulation of the Voltage-Gated Ca^{2+} Current by Intracellular Free Mg^{2+} Studied by Internal Perfusion of Single Cardiac Myocytes

H. Criss Hartzell[1] and Richard E. White[1,2]

Introduction

Despite the fact that Mg^{2+} is the major divalent cation in the cell, relatively little is known about the regulation of its intracellular concentration or the effects of changes in this on cellular function. Recently, it has become evident that a variety of different kinds of ion channels are affected by intracellular Mg^{2+}. For example, inward rectification of several kinds of K^+ channels including the inward rectifier, the ATP-dependent K^+ channel, and the acetylcholine (ACh)-activated K^+ channel has been shown to be caused by a time- and voltage-dependent blockade of the K^+ channel by intracellular Mg^{2+} (Matsuda et al. 1987; Matsuda 1988; Horie and Irisawa 1987, 1989; Horie et al. 1987; Vandenberg 1987). There are reasons to believe, however, that the effects of Mg^{2+} on ionic channels are not limited to simple obstruction of the permeation pathway. For example, the delayed rectifier current (I_K) is reduced in amplitude by increasing intracellular Mg^{2+}, but the mechanism of this effect is rather insensitive to voltage, suggesting that Mg^{2+} blocks this channel at a site that is remote from the pore itself (Duchatelle-Gourdon et al. 1989). Recently, Pusch et al. (1989) have shown that outward current through tetrodotoxin (TTX)-sensitive Na^+ channels is blocked in a voltage- and time-dependent fashion by intracellular Mg^{2+} in a manner consistent with Mg^{2+} blockage of ion permeation. However, in addition, internal Mg^{2+} also affects inward current through these channels, implying another site of Mg^{2+} action.

In this article, we will summarize work in my lab (White and Hartzell 1988, 1989; Hartzell and White 1989) on the modulation of voltage-gated Ca^{2+} currents by intracellular Mg^{2+}. Voltage-gated Ca^{2+} channels are present in most, if not all, excitable cells and play a key role in the initiation and maintenance of contraction of cardiac cells (Hartzell 1988). Thus, understanding the mechanisms by which these channels are affected by intracellular constituents such as Mg^{2+} is an important step in understanding how these channels are regulated physiologically and how their malfunction may occur pathologically.

[1] Department of Anatomy and Cell Biology, Emory University School of Medicine, Atlanta, GA 30322, USA
[2] Present address: NIEHS, Laboratory of Cellular and Molecular Pharmacology, Research Triangle Park, NC 27709, USA

Complex effects of Mg^{2+} have been found on I_{Ca} amplitude and inactivation that can be partly explained by supposing that Mg^{2+} binds to site in the permeation pathway, thus obstructing the passage of permeant ions and also altering the mode of channel inactivation. Furthermore, the binding of Mg^{2+} to this site is modulated by channel phosphorylation.

Methods

Single myocytes were isolated from frog ventricle using enzymatic and mechanical dissociation procedure that have been described in detail elsewhere (Fischmeister and Hartzell 1987; White and Hartzell 1988; Hartzell and White 1989). I_{Ca} was measured using the whole-cell configuration of the patch clamp technique (Hamill et al. 1981). I_{Na} was blocked with TTX. K^+ currents were blocked by both internal and external Cs^+. Sometimes internal tetraethylammonium (TEA) was also used. Currents were elicited by 200- or 400-ms duration voltage clamp pulses. They were leak-subtracted using the following procedure. Leak was determined by averaging several current traces evoked by voltage pulses from −80 mV to −70 mV. The mean holding current and leak current were calculated from the averaged trace and used to construct an idealized noise-free leak trace. The trace was then multiplied by the appropriate scaling factor, assuming a linear current–voltage (I–V) relationship for the leak, and subtracted from the I_{Ca}, I_{Ba}, or $I_{Ca(Na)}$ traces. The voltage protocols for determining I_{Ca} inactivation are based on the two-pulse paradigm of Hodgkin and Huxley (1952) and have already been published (Fischmeister and Hartzell 1986; Argibay et al. 1988). Half-inactivation was defined as the prepulse potential producing 50% of the inactivation produced by the same duration prepulse to 0 mV.

External solutions contained 115 m*M* NaCl, 20 m*M* CsCl, 0.3–3.0 m*M* $MgCl_2$, 1.8 m*M* $CaCl_2$ or 2 m*M* $BaCl_2$, 5 m*M* sodium pyruvate, 5 m*M* glucose, 0.001 m*M* TTX, and 10 m*M* HEPES (*N*-2-hydroxyethyl-piperazine-*N'*-2-ethane sulfonic acid), at pH 7.4. External Mg^{2+} concentration was varied between 0.3 m*M* and 3 m*M* without compensation for changes in osmotic strength. When Ba^{2+} was used as charge carrier, Ca^{2+} was absent. When Na^+ was the charge carrier, the external solution contained 25 m*M* NaCl, 110 m*M* CsCl, 0.1 m*M* EGTA, and 10 m*M* HEPES at pH 8.5. The internal solution varied in composition and could be changed during recording by a system permitting continuous perfusion of the patch electrode (Hartzell and Fischmeister 1986; Fischmeister and Hartzell 1987).

The goal of these experiments was to examine the effects of changing free intracellular Mg^{2+} concentration while maintaining MgATP and other ion concentrations as constant as possible. Composition of the solutions was calculated by a computer program developed by Godt and Lindley (1982), which was kindly provided by Dr. Robert Godt (Medical College of Georgia, Augusta). The stability constants were the same as those published by Godt and Lindley (1982). In these calculations, I assumed that Cs^+ was equivalent to K^+. Solution composition is given in more detail in Hartzell and White (1989).

Currents were fitted to sums of exponentials by a Marquart least squares procedure (FIT program, W. Goolsby, Emory University). Leak-subtracted currents were used for the least squares fitting routine, and the current between the peak inward current and the end of the pulse was subjected to the curve-fitting routine.

Results

Effects of Mg^{2+} on Basal I_{Ca} and I_{Ba} Amplitude

The effect of internal Mg^{2+} on I_{Ca} was examined by eliciting I_{Ca} with 400-ms voltage pulses from -80 mV holding potential to various potentials at 8-s intervals. $[Mg^{2+}]_i$ was altered by changing the solution in the recording pipette by means of a perfusion system. There were significant differences in the effect of $[Mg^{2+}]_i$ depending on the charge carrier. At the peak of the I–V relationship, the effects of $[Mg^{2+}]_i$ on I_{Ca} amplitude were minimal, whereas the effects on Ba^{2+} (I_{Ba}) or Na^+ ($I_{Ca(Na)}$) current through Ca^{2+} channels were greater. Specifically, in 15 cells that were perfused with both 0.3 m*M* Mg^{2+} and 3.0 m*M* Mg^{2+}, I_{Ca} at 0 mV was reduced $11\% \pm 8.5\%$ (mean $\pm$ SE) upon switching to the higher Mg^{2+} concentration. The largest reduction observed was 34%. In comparing two partially overlapping populations of cells that were perfused with either 0.3 m*M* or 3 m*M* Mg^{2+} or both concentrations, I_{Ca} in the presence of 3 m*M* Mg_i^{2+} was 20% less than it was in the presence of 0.3 m*M* Mg_i^{2+} ($n=97$ for 0.3 m*M* and $n=45$ for 3 m*M* Mg^{2+}). The effects of internal Mg^{2+} on the amplitude of I_{Ba} were somewhat greater: I_{Ba} in the presence of 3.0 m*M* Mg^{2+} was reduced 36% compared with I_{Ba} in the presence of 0.3 m*M* Mg^{2+} in 25 cells studied. In approximately 10 cells in which we used Na^+ as the charge carrier in the absence of any external divalent cations, increasing $[Mg^{2+}]_i$ caused a very significant decrease in $I_{Ca(Na)}$: at -20 mV it was reduced by 50%–80%.

In addition to producing these changes in maximal current amplitude, increasing $[Mg^{2+}]_i$ also shited the I–V relationships to more negative potentials (Fig. 1). The magnitude of this effect was dependent upon the charge carrier. The I–V relationship for I_{Ca} was shifted an average of 7 mV to the left, that for I_{Ba}, 10 mV, and that for $I_{Ca(Na)}$, 20 mV. Thus, some of the decrease in the currents observed at 0 mV is due to the shift in the I–V relationship. Despite this shift, however, the maximum current in the presence of 3 m*M* Mg^{2+} was smaller than in the presence of 0.3 m*M* Mg^{2+} when Ba^{2+} or Na^+ was the charge carrier.

We can suggest two possible mechanisms by which the shift in the I–V relationship might be produced by $[Mg^{2+}]_i$. $[Mg^{2+}]_i$ could shift the I–V relationship by screening a negative internal surface charge. Such a mechanism would be nonspecific and would not involve any direct effect on Ca^{2+} chan-

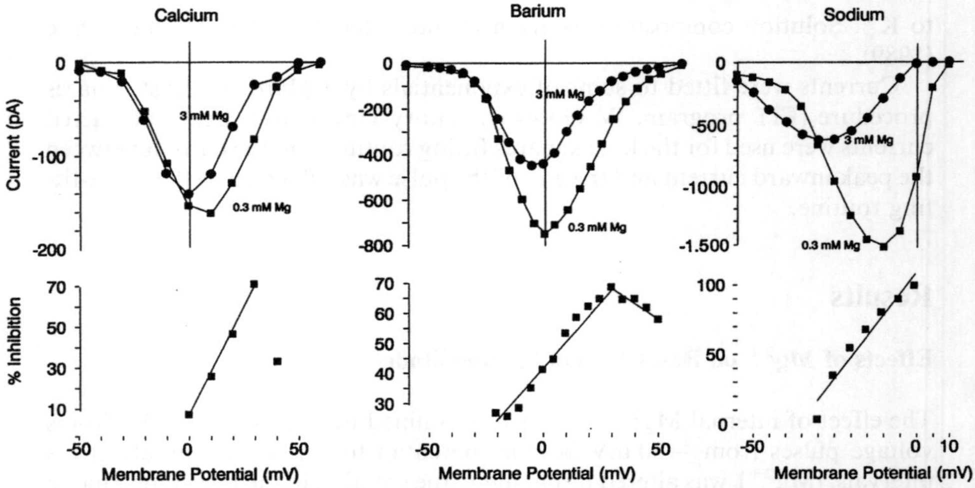

Fig. 1. Current–voltage relationships of I_{Ca}, I_{Ba}, and $I_{Ca(Na)}$ with different $[Mg^{2+}]_i$. Current–voltage relationships were obtained by 400-ms pulses from −80 mV holding potential. Currents were leak subtracted by subtracting a linearly scaled current produced in response to a 10-mV depolarization. *Solid squares*, internal solution of 0.3 m*M* Mg^{2+}. *Solid circles*, internal solution of 3 m*M* Mg^{2+}. Permeant ion species were: 1.8 m*M* Ca, 5 m*M* Ba, 25 m*M* Na. Solutions are described in detail in Methods. *Lower portion* shows the percentage inhibition of inward current as a function of membrane potential produced by switching from 0.3 m*M* Mg^{2+} to 3 m*M* Mg^{2+}

nels. Alternatively, $[Mg^{2+}]_i$ could compete for permeant ion binding to permeation sites in the channel in a voltage-dependent manner. These possibilities are explored in more detail below.

Effects of Mg^{2+} on I_{Ca} and I_{Ba} Inactivation Under Basal Conditions

Ca^{2+} Channel Inactivation

I_{Ca} inactivation occurs by two mechanisms: one is voltage dependent and the other is Ca^{2+} dependent. In general (Eckert and Chadd 1984), these two mechanisms can be distinguished experimentally by using Ba^{2+} or Na^+ as the charge carrier, because these ions do not activate the Ca^{2+}-dependent component of inactivation (Lee et al. 1985), or by buffering internal Ca^{2+} to low levels with EGTA or BAPTA (Bechem and Pott 1985). In our studies, however, using Ba^{2+} as the charge carrier slows inactivation much more than does Ca^{2+} buffering (Argibay et al. 1988). Even with 25 m*M* BAPTA internally, I_{Ca} inactivates more rapidly than does I_{Ba} (H.C. Hartzell, unpublished data). This suggests that either it is impossible to buffer Ca^{2+} at the site responsible for Ca^{2+}-dependent inactivation, perhaps because this site is in the permeation pathway, or that Ba^{2+} affects channel gating independently of Ca^{2+}-dependent inactivation. This possibility is strengthened by the observation that the inactivation kinetics of I_{Ba} depend upon the Ba^{2+} concentration, even in the nominal absence of Ca^{2+}. Ignoring these possible complications, the effects of $[Mg^{2+}]_i$ on I_{Ca} and I_{Ba} inactivation were examined in order to gain insights into the mechanism of the Mg^{2+} effect.

Before looking at the effects of $[Mg^{2+}]_i$ on inactivation, the inactivation of I_{Ca} and I_{Ba} was characterized. The rates of Ca^{2+}-dependent and voltage-dependent inactivation were estimated by a standard gapped two-pulse protocol consisting of a prepulse of varying duration and amplitude followed after a 3-ms interval by a 200-ms test pulse from −80 mV to 0 mV. The amplitude of the current elicited by the test pulse was expressed as a percentage of the current in the absence of a prepulse. Ca^{2+} buffering in these experiments was moderate: The internal solution contained only 5 m*M* EGTA.

The I_{Ca} inactivation curve (Fig. 2A) had a "U"-shape as described previously by others (Tillotson 1979; Eckert and Tillotson 1981; Mentrard et al. 1984; Hadley and Hume 1987; Argibay et al. 1988). Increasing inactivation of I_{Ca} occurred with prepulses between −60 mV and 0 mV, with maximum inactivation occurring as prepulses neared 0 mV. With prepulses of short duration, inactivation decreased as prepulses became positive to 0 mV. This "relief from inactivation" (Lee et al. 1985) is due to the fact that the inactivation curve with Ca^{2+} as the charge carrier is the sum of voltage-dependent inactivation, which is incomplete especially with short duration prepulses, and Ca^{2+}-dependent inactivation, which is greatest in magnitude at the peak of the I−V relationship (~0 mV). The inactivation curves obtained in the same cell using Ba^{2+} as the charge carrier reflect the voltage-dependent component of inactivation (Fig. 2B). Under these conditions, the inactivation curve was much less U-shaped, and relief from inactivation was usually negligible. The curve was sigmoidal with a plateau between 0 and +100 mV.

The time course of inactivation was evaluated by examining the effect of different duration prepulses to 0 mV on the response to the standard test pulse (Fig. 3A). The time course of voltage-dependent inactivation with Ba^{2+} as the charge carrier was slower than inactivation produced by Ca^{2+}-dependent and voltage-dependent processes together with Ca^{2+} as the charge carrier. With Ba^{2+} as the charge carrier, 50% inactivation was produced by 60-ms prepulses, and steady-state inactivation was achieved with 700-ms prepulses. With Ca^{2+} as the charge carrier, inactivation was more than twice as fast and was complete within 200 ms. The amount of Ca^{2+}-dependent inactivation relative to the voltage-dependent component was calculated as the ratio of the difference between the two curves and the Ba^{2+} curve (Fig. 3B). Ca^{2+}-dependent inactivation contributed most to inactivation with short prepulses and declined as the amount of voltage-dependent inactivation increased.

Effects of $[Mg^{2+}]_i$ on Inactivation Kinetics

$[Mg^{2+}]_i$ had only a small effect on the kinetics of inactivation of I_{Ca} at any potential (data for 0 mV shown in Fig. 4). I_{Ca} decayed almost completely

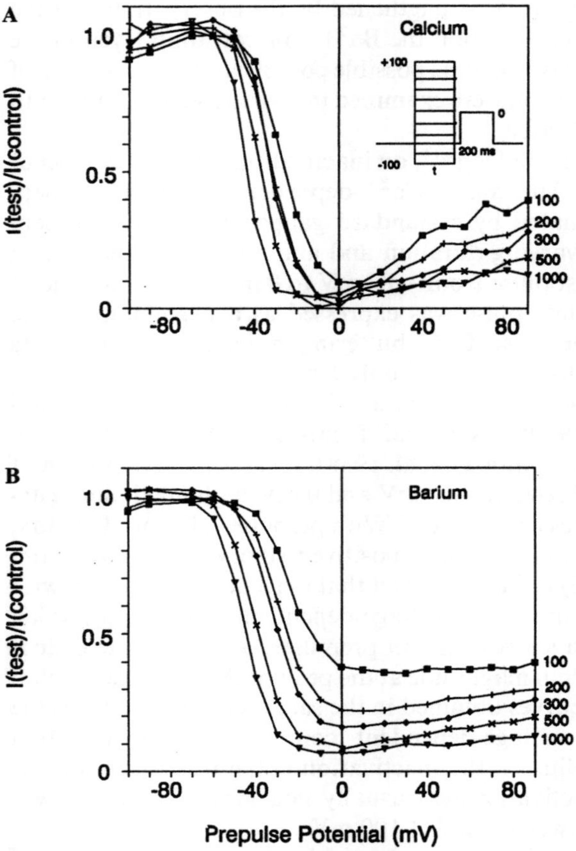

Fig. 2 A, B. Inactivation of I_{Ca} and I_{Ba}. **A** I_{Ca} inactivation. Internal solution contained 1.0 m*M* Mg^{2+}. External solution contained 1.8 m*M* $CaCl_2$, 1.8 m*M* $MgCl_2$. A two-pulse inactivation protocol consisted of a prepulse to the voltage indicated on the X-axis of duration 100 ms (*square*), 200 ms (*plus*), 300 ms (*diamond*), 500 ms (*cross*), or 1000 ms (*inverted triangles*) followed after a 3-ms interval by a test pulse to 0 mV of 200-ms duration. Amplitude of response to the test pulse preceded by a prepulse was plotted as a fraction of the amplitude of the response to the test pulse in the absence of a prepulse. **B** I_{Ba} inactivation from the same cells as in **A.** Solutions and conditions were the same as in **A** except that 2 m*M* $BaCl_2$ replaced 1.8 m*M* $CaCl_2$ in the external solution. Reproduced from the *Journal of General Physiology* (1989) by copyright permission of The Rockefeller University Press (Hartzell and White 1989)

within 200 ms with both 0.3 m*M* and 3.0 m*M* Mg_i^{2+}. Although a quantitative analysis of Ca^{2+} current kinetics in different $[Mg^{2+}]_i$ remains to be accomplished, in many cells increasing $[Mg^{2+}]_i$ caused a slight slowing of I_{Ca} (Fig. 4). This is more evident in the lower set of traces in Fig. 4, which were obtained in the presence of a low concentration of cAMP, but was also true in the absence of cAMP (upper set of traces).

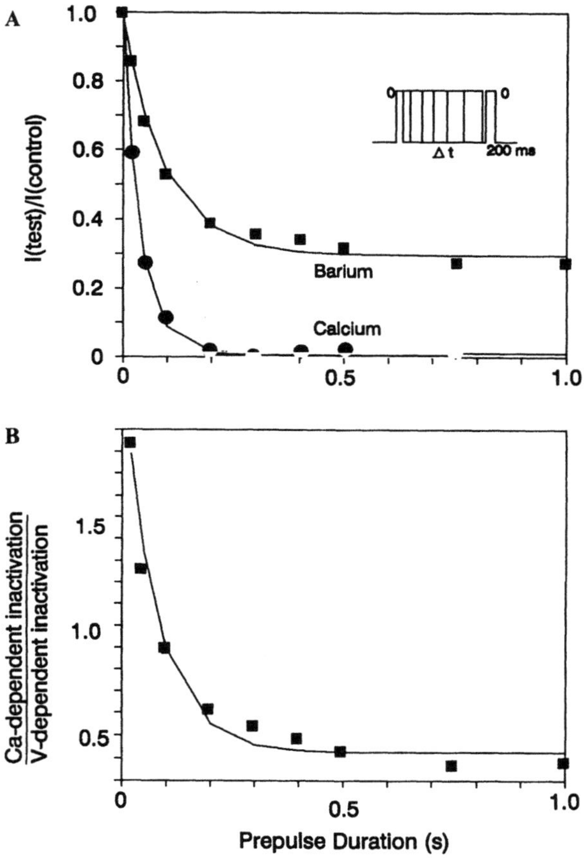

Fig. 3. A Time course of inactivation of I_{Ca} and I_{Ba} at 0 mV. A two-pulse protocol consisted of prepulses to 0 mV for the durations shown on the X-axis followed after a 3-ms interval by a 200-ms test pulse to 0 mV. Inactivation was plotted as in Fig. 2. Solutions and conditions were the same as in Fig. 2. *Square,* 2 m*M* $BaCl_2$ in external solution; *circles,* 1.8 m*M* $CaCl_2$ in external solution. **B** Time course of Ca^{2+}-dependent inactivation relative to voltage-dependent inactivation at 0 mV. Ca^{2+}-dependent inactivation was defined as the difference between the inactivation curves in Ba^{2+} and in Ca^{2+}. Voltage-dependent inactivation was defined as the curve in Ba^{2+}. Ratio between Ca^{2+}-dependent and voltage-dependent inactivation is plotted as a function of the prepulse duration. Reproduced from the *Journal of General Physiology* (1989) by copyright permission of The Rockefeller University Press (Hartzell and White 1989)

In contrast to the small effect of $[Mg^{2+}]_i$ on I_{Ca} kinetics, increasing $[Mg^{2+}]_i$ altered the inactivation of I_{Ba} and $I_{Ca(Na)}$ drastically (Fig. 5). With 0.3 m*M* Mg_i^{2+}, I_{Ba} at 0 mV decayed less than 50% in 200 ms and was still incompletely inactivated after 800 ms. Changing $[Mg^{2+}]_i$ to 3.0 m*M* produced a marked increase in I_{Ba} decay such that the current was almost completely inactivated within 200 ms. The kinetics of inactivation of I_{Ba} were altered at all potentials.

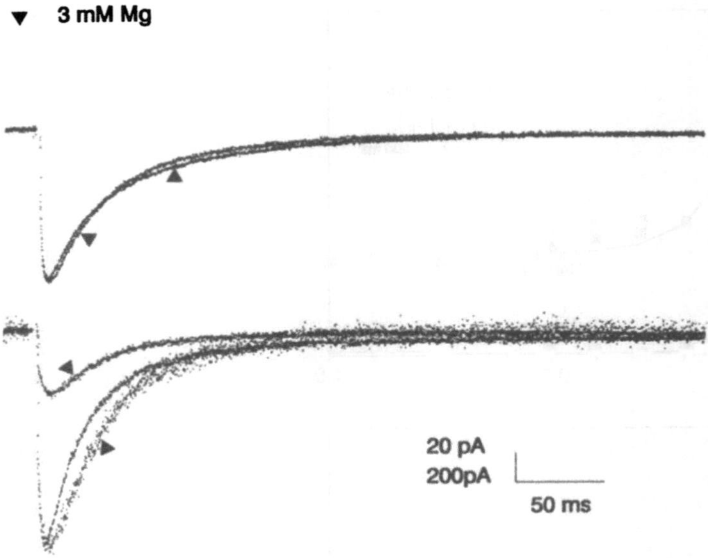

Fig. 4. Effect of $[Mg^{2+}]_i$ on I_{Ca} kinetics. Results from two different cells are shown to show the range of effect. *Upper panel,* two current traces are superimposed. That indicated by *triangles* was obtained in 3 m*M* $[Mg^{2+}]_i$, and the *unmarked trace* was obtained in 0.3 m*M* $[Mg^{2+}]_i$. *Lower panel,* three traces are superimposed, which were obtained from a cell internally perfused with 1 m*M* cAMP. *Smallest trace* was obtained in 3 m*M* Mg_i^{2+}. It was then normalized to the same amplitude as a trace obtained in 0.3 m*M* Mg_i^{2+} (*unmarked trace*). In both panels, to different extents, I_{Ca} obtained in 3 m*M* Mg_i^{2+} decayed more slowly than that in 0.3 m*M* Mg_i^{2+}

Although the effect of $[Mg^{2+}]_i$ on $I_{Ca(Na)}$ has not been analyzed as extensively, the results appear qualitatively the same as those obtained with I_{Ba}. $I_{Ca(Na)}$ obtained at all potentials inactivated more rapidly with 3 m*M* Mg_i^{2+} than with 0.3 m*M* Mg_i^{2+}.

A quantitative analysis of the kinetics of inactivation of I_{Ba} is shown in Fig. 6. Ba^{2+} currents were leak subtracted and fitted to the sum of two exponentials plus an offset ($I_{Ba} = A * \exp(-t/\tau_2) + B * \exp(-t/\tau_2) + C$). Each panel in Fig. 6 is a 3-dimensional plot showing the amplitudes (A, B, and C; Z-axis) and the time constants (τ_1 and τ_2; Y-axis) of each component as a function of test potential. Arbitrarily, the fast component was defined as having a time constant of less than 100 ms. The most noticeable difference

→

Fig. 5. Effect of $[Mg^{2+}]_i$ on the kinetics of I_{Ba} and $I_{Ca(Na)}$. Current traces from experiments similar to those described in Fig. 1 are shown. *Left traces* were obtained with 0.3 m*M* Mg_i^{2+} and *right traces* with 3 m*M* Mg_i^{2+}. Ba^{2+} and Na^+ experiments were from different cells. I_{Ba} was evoked by 800-ms duration pulses from -80 mV to the potential indicated. Traces were leak subtracted. External Mg^{2+} concentration was 0 m*M*. $I_{Ca(Na)}$ was obtained with 400-ms duration pulses from -80 mV to the potential indicated. Traces were leak-subtracted

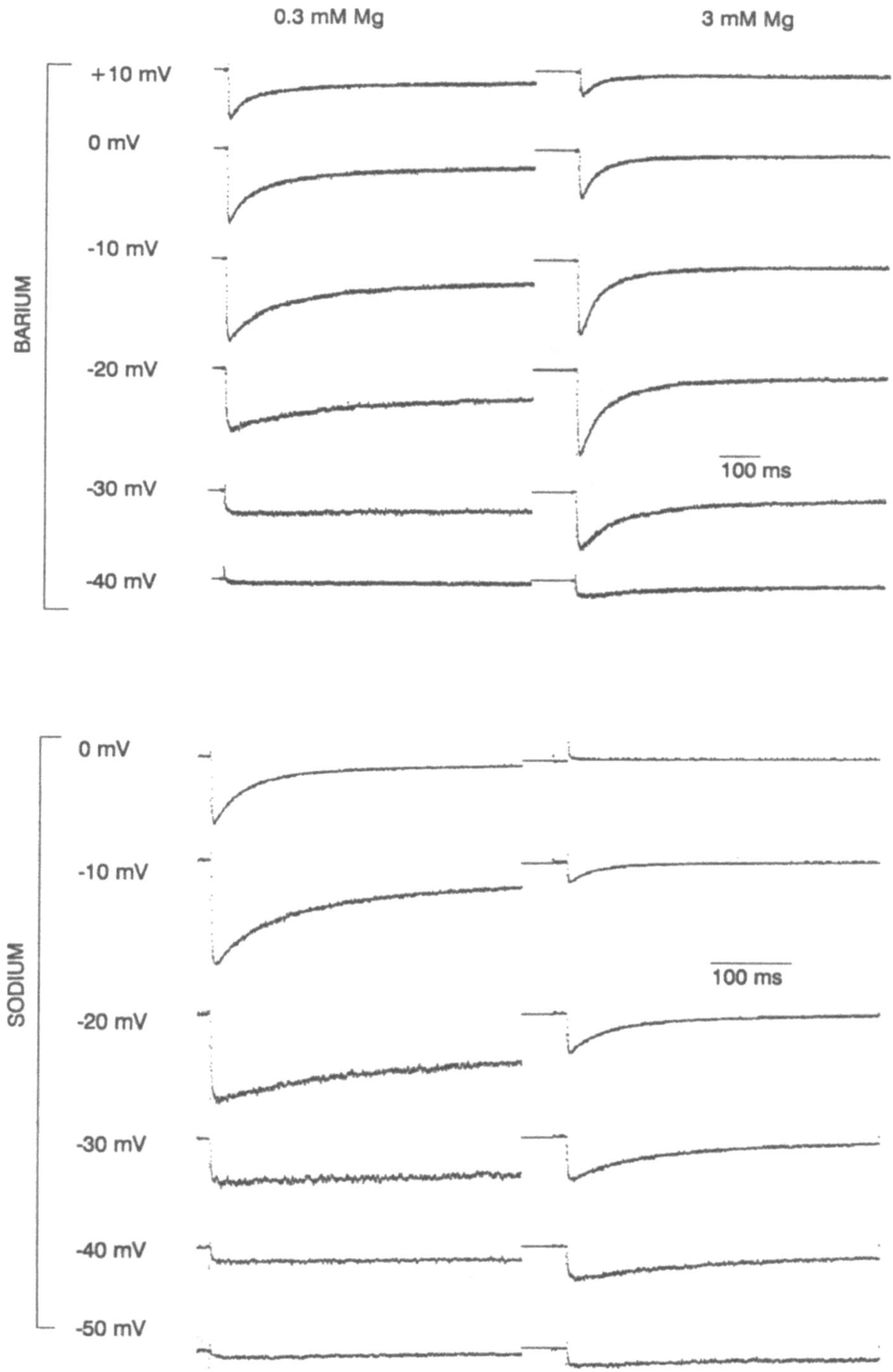
0.3 mM Mg
3 mM Mg
BARIUM
+10 mV
0 mV
-10 mV
-20 mV
-30 mV
-40 mV
100 ms
SODIUM
0 mV
-10 mV
-20 mV
-30 mV
-40 mV
-50 mV
100 ms

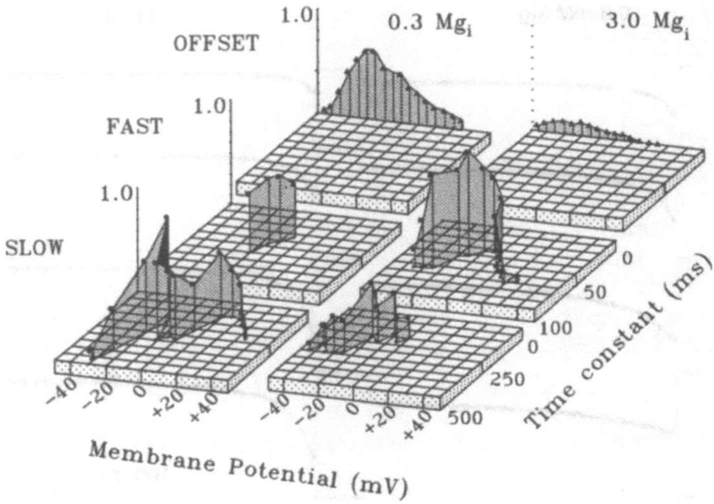

Fig. 6. Time constants of fits of I_{Ba}. Ba^{2+} currents were fitted to the sum of two exponentials plus an offset (see text). Amplitudes were normalized as a fraction of the largest amplitude component recorded for each Mg^{2+} concentration (this was the slow component at -15 mV for 0.3 m*M* Mg_i^{2+} and the fast component at -15 mV for 3.0 m*M* Mg_i^{2+}). Amplitude (z-axis) and time constant (y-axis) of each component are plotted as a function of test pulse voltage. Reproduced from the *Journal of General Physiology* (1989) by copyright permission of The Rockefeller University Press (Hartzell and White 1989)

between the currents obtained in 0.3 m*M* and 3.0 m*M* Mg_i^{2+} was the relative magnitude of the offset. The offset was several-fold larger at almost every potential in 0.3 m*M* than it was in 3.0 m*M* Mg_i^{2+}. The rapidly inactivating component was the major component with 3.0 m*M* Mg_i^{2+} at all voltages, but with 0.3 m*M* Mg_i^{2+} the rapidly inactivating component was a minor component between $+5$ mV and -10 mV and was negligible at all other potentials. The fast time constant was similar with both 0.3 m*M* a 3.0 m*M* Mg_i^{2+} (between 30 and 60 ms), but the slow time constant was considerably slower in 0.3 m*M* Mg_i^{2+} than in 3.0 m*M* Mg_i^{2+}. Thus, these data show that increasing $[Mg^{2+}]_i$ increased the relative magnitude of the noninactivating component of I_{Ba} and slowed the time constant and increased the amplitude of the slow component at all potentials. Thus, it appears that Mg_i^{2+} causes an increase in the voltage-dependent component of inactivation.

The effect of high $[Mg^{2+}]_i$ on the amplitude and decay of I_{Ba} was at least partially reversible except after prolonged (>30 min) internal perfusion. However, it was seldom that we were able to reverse completely the effects of high $[Mg^{2+}]$. This irreversibility was associated with a gradual contracture of the cell with repeated stimulation. We supposed that the contracture was related to an accumulation of intracellular Ba^{2+}, because Ba^{2+} can activate the contractile machinery but is not chelated well by EGTA and is not pumped by most Ca^{2+}-ATPases, including the sarcolemmal Ca^{2+}-ATPase and the sarcoplasmic reticulum Ca^{2+}-pump. Accumulation of intracellular Ba^{2+}, however, does not explain the changes in inactivation kinetics, because increases in the inactivation rate of I_{Ba} were also observed when the cells did not undergo contracture. Several experiments were performed with a very low stimulation frequency (1 min^{-1} or less) to reduce Ba^{2+} influx. These cells did not visibly contract, but I_{Ba} inactivation was nevertheless increased by increasing $[Mg^{2+}]_i$. Furthermore, as shown above, similar changes in inactivation kinetics were observed when Na^+ was used as the charge carrier. The possibility should be recognized, however, that the effects of $[Mg^{2+}]_i$ on inactivation could be due to complex alterations in cellular ionic homeostasis. For example, Mg^{2+}, Ba^{2+}, Ca^{2+}, and Na^+ likely compete for binding to various sites in the cell. Maintenance of homeostasis of divalent cations also involves various co-transport mechanisms involving monovalent cations. Thus, altering one ion could result in complex changes in ionic composition in the cell. Although specific, testable hypotheses may not immediately come to mind, these problems are worthy of causing the loss of a considerable amount of sleep.

Effects of $[Mg^{2+}]_i$ on Quasi-Steady-State Inactivation

$[Mg^{2+}]_i$ had characteristic effects on inactivation as measured by the gapped two-pulse protocol with 200-ms duration prepulses (Fig. 7). With Ca^{2+} as the charge carrier, increasing $[Mg^{2+}]_i$ shifted half-inactivation about 5 mV in the negative direction and depressed relief from inactivation considerably. Relief from inactivation with prepulses to +100 mV decreased from 85% to 30% as $[Mg^{2+}]_i$ was increased from 0.3 m*M*. On average, with Ca^{2+} as the charge carrier, increasing $[Mg^{2+}]_i$ from 0.3 m*M* to 3.0 m*M* produced a 56% depression in relief from inactivation. With Ba^{2+} as the charge carrier, half-inactivation was shifted 8 mV in a negative direction, and the plateau of inactivation between 0 and +100 mV was diminished by more than 50% when $[Mg^{2+}]_i$ was increased. These results are consistent with the hypothesis that internal Mg^{2+} increases the amount of voltage-dependent inactivation.

Since the inactivation protocol with 200-ms prepulses did not produce steady-state inactivation of I_{Ba}, the difference in the inactivation curves with different $[Mg^{2+}]_i$ could have been caused by changes in the rate of inactivation as well as in the level of steady-state inactivation. The results above show clearly that the rate of inactivation is increased by increasing $[Mg^{2+}]_i$. In addition, analysis of the rate of steady-state inactivation measured by the standard two-pulse protocol suggests that steady-state inactivation occurs at least twice as rapidly with 0.3 m*M* Mg_i^{2+} than with 3.0 m*M* Mg_i^{2+}.

In addition to its effect on the rate of inactivation $[Mg^{2+}]_i$ also increases the steady-state level of inactivation. In the cell shown in Fig. 8, I_{Ba} inactivation was measured with the two-pulse protocol using 3-s prepulses. When the cell was perfused with 3.0 m*M* Mg_i^{2+}, half-inactivation was shifted in the negative direction, and the steady-state level of inactivation was considerably less.

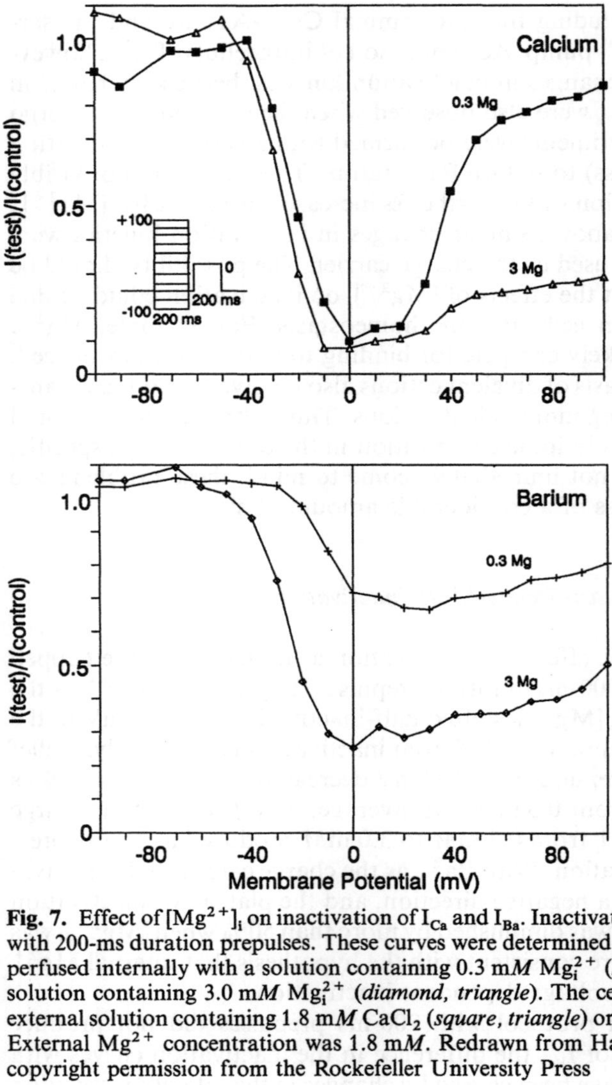

Fig. 7. Effect of $[Mg^{2+}]_i$ on inactivation of I_{Ca} and I_{Ba}. Inactivation was measured as in Fig. 2 with 200-ms duration prepulses. These curves were determined in the same cell that was first perfused internally with a solution containing 0.3 m*M* Mg_i^{2+} (*square, plus*), and then with a solution containing 3.0 m*M* Mg_i^{2+} (*diamond, triangle*). The cell was alternately exposed to external solution containing 1.8 m*M* $CaCl_2$ (*square, triangle*) or 2 m*M* $BaCl_2$ (*diamond, plus*). External Mg^{2+} concentration was 1.8 m*M*. Redrawn from Hartzell and White (1989) with copyright permission from the Rockefeller University Press

These results suggest that voltage-dependent inactivation is increased by $[Mg^{2+}]_i$. In order to test whether voltage-dependent inactivation was due entirely to a Mg^{2+}-dependent mechanism, we examined the kinetics of I_{Ba} in cells exposed both internally and externally to Mg^{2+}-free solutions (data not shown, see Hartzell and White 1989). The internal solution was similar to the 0.3 m*M* Mg^{2+} solution except that $MgCl_2$ and ATP were omitted and EDTA was added. Under these conditions I_{Ba} was slower than that observed in 0.3 m*M* Mg_i^{2+}, but the current still exhibited both an inactivating component

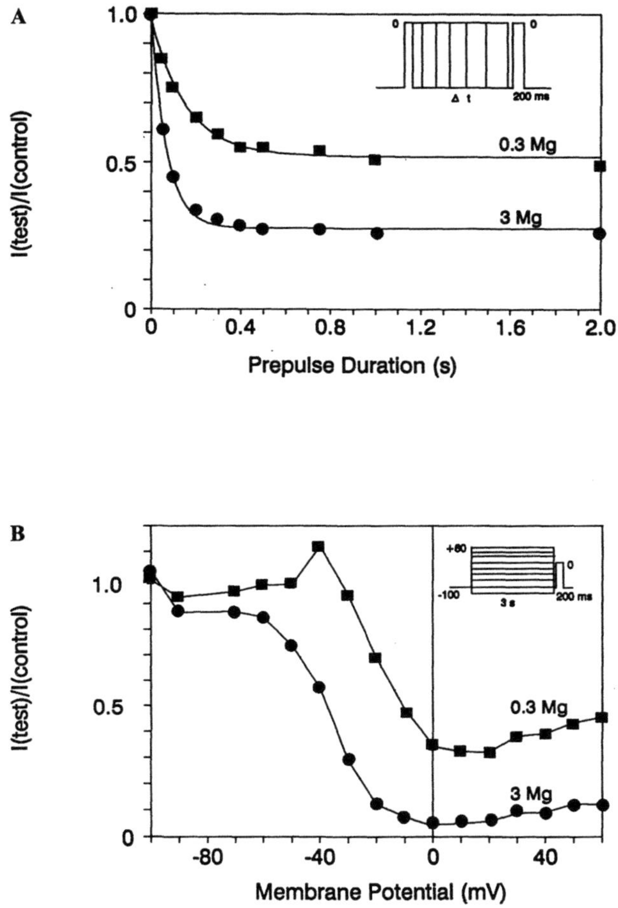

Fig. 8 A, B. Effect of $[Mg^{2+}]_i$ on steady-state inactivation of I_{Ba}. **A** Time course of inactivation of I_{Ba} measured with a two-pulse protocol. Inactivation was determined as in Fig. 3 in the presence of either 0.3 m*M* or 3.0 m*M* Mg_i^{2+}. **B** Inactivation was determined as in Fig. 2 with 3-s duration prepulses. Cell was first perfused internally with 0.3 m*M* Mg_i^{2+} (*square*) followed by 3.0 m*M* Mg_i^{2+} (*circles*). Reproduced from the *Journal of General Physiology* (1989) by copyright permission of The Rockefeller University Press (Hartzell and White 1989)

and a non-inactivating component. Thus, either voltage-dependent inactivation is caused entirely by a Mg^{2+}-dependent mechanism, or it is not possible to buffer intracellular Mg^{2+} completely by this method.

The effects of $[Mg^{2+}]$ were specific to the side of the membrane on which the $[Mg^{2+}]$ was manipulated. Changing $[Mg^{2+}]_o$ altered I_{Ba} amplitude, as has been reported previously, but had little or no effect on I_{Ba} or I_{Ca} inactivation. Increasing the external Mg^{2+} concentration had effects on the kinetics of inactivation that were qualitatively the opposite to those of increasing $[Mg^{2+}]_i$

except that these effects were relatively small. Specifically, increasing $[Mg^{2+}]_o$ increased the time constant and amplitude of the slow component of inactivation and may have increased the offset slightly. Furthermore, the inactivation curves measured with 200-ms prepulses with Ca^{2+} as the charge carrier have the same U-shape with different external Mg^{2+} concentrations (data not shown, but see Hartzell and White 1989).

Frog Ventricular Cells Exhibit Only L-type Ca^{2+} Channels

The ability of internal Mg^{2+} to convert a slowly inactivating current to a rapidly inactivating one suggested the possibility that internal Mg^{2+} was converting or altering the expression of T- and L-type Ca^{2+} channels (Bean 1985): High $[Mg^{2+}]_i$ might favor the expression of T-type channels. However, we were unable to find any evidence for T-type channels in this preparation using different holding potentials or drugs including dihydropyridine Ca^{2+} channel antagonists, nickel, or tetramethryn. With 0.3 m*M* Mg_i^{2+} (Fig. 9 A) the I–V relationships of I_{Ba} from holding potentials of −100 mV and −40 mV and the difference between these curves were virtually identical in shape, and all had peaks near 0 mV. I_{Ba} was completely blocked by 0.1–1.0 μ*M* nifedipine at holding potentials positive to −50 mV and was blocked 80% at a holding potential of −100 mV. This incomplete block was most likely due to the voltage-dependent nature of the nifedipine block (Kass and Krafte 1987). With 3.0 m*M* Mg_i^{2+} (Fig. 9 B), I_{Ba} was almost completely inactivated at a holding potential of −40 mV, as expected from the effects of Mg^{2+} on the steady-state inactivation curve. The I–V relationship from a holding potential of −50 mV, however, had the same shape as that obtained from a −100 mV holding potential. I_{Ba} was completely blocked by 1 μ*M* nifedipine even from a holding potential of −100 mV under these conditions. Thus, there was no evidence for T-type Ca^{2+} channels in this preparation.

Effects of $[Mg^{2+}]_i$ on I_{Ca} and I_{Ba} Elevated by cAMP-Dependent Mechanisms

In contrast to the small effect of $[Mg^{2+}]_i$ on the amplitude of basal I_{Ca} at 0 mV, $[Mg^{2+}]_i$ had a much larger effect on the amplitude of I_{Ca} elevated by cAMP-dependent mechanisms. This was true when cAMP levels were raised by β-adrenergic agonists, by forskolin, or by internal perfusion of the cell with cAMP. In Fig. 10, the top panel shows the amplitude of I_{Ca} measured near the peak of the I–V relationship (0 mV) as a function of time upon changing from 0.3 m*M* Mg_i^{2+} to 3 m*M* Mg_i^{2+}. On the average, I_{Ca} at 0 mV decreased 65% in response to the increase in $[Mg^{2+}]_i$. However, as was observed for basal I_{Ca}, the effect of $[Mg^{2+}]_i$ is voltage dependent (Fig. 10, middle panel), and I_{Ca} is reduced more at more positive potentials (Fig. 10, bottom panel).

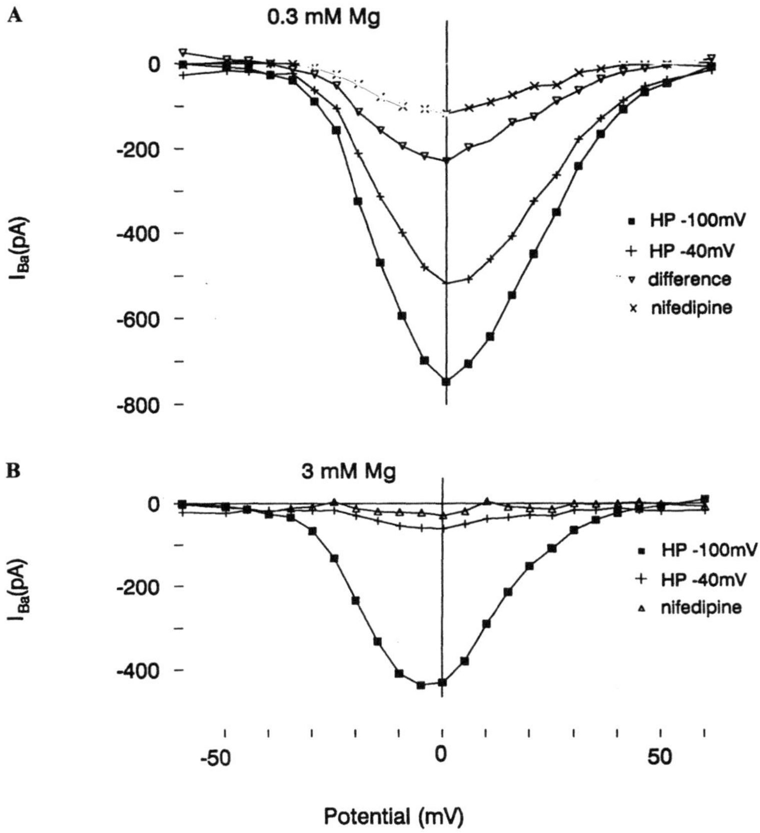

Fig. 9 A, B. Absence of T-type Ca^{2+} channels in frog ventricular cells. **A** Current–voltage relationships obtained in 0.3 m*M* $[Mg^{2+}]_i$ from holding potentials of −100 mV (*square*) and −40 mV (*plus*). The difference between these curves is shown as *inverted triangles*. I–V relationship obtained in the presence of 1.0 μ*M* nifedipine with a −100 mV holding potential is shown as *cross*. **B** I–V relationships obtained in 3.0 m*M* Mg_i^{2+} under the same conditions. Reproduced from the *Journal of General Physiology* (1989) by copyright permission of The Rockefeller University Press (Hartzell and White 1989)

Mechanisms of the Effect of Mg^{2+} on cAMP-Elevated I_{Ca}

The effect of $[Mg^{2+}]_i$ on the amplitude of I_{Ca} elevated by cAMP-dependent mechanisms could be due to effects of $[Mg^{2+}]_i$ on cAMP production, cAMP degradation, or phosphorylation of the Ca^{2+} channel. Since I_{Ca} was reduced by increasing $[Mg^{2+}]_i$ even when the cell was perfused internally with cAMP (Fig. 10), this suggested that $[Mg^{2+}]_i$ was not interfering with cAMP synthesis. In order to test whether the effects of cAMP observed were related to these mechanisms, we examined the effect of $[Mg^{2+}]_i$ on I_{Ca} elevated by various agents that bypass various steps in the cAMP-dependent phosphorylation

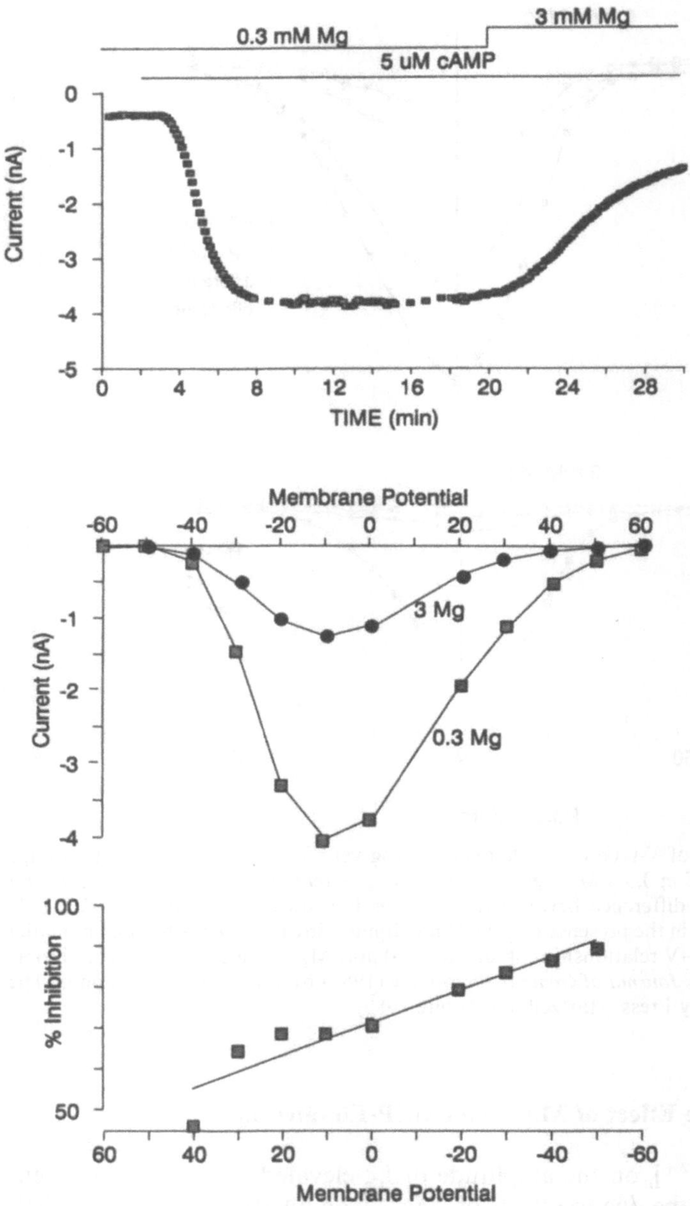

Fig. 10. Effects of changing $[Mg^{2+}]_i$ on I_{Ca} elevated by internal perfusion with 5 μ*M* cAMP. *Top panel,* cell was initially perfused with 0.3 m*M* Mg_i^{2+}, and 5 μ*M* cAMP was added to the perfusion where indicated (t = 2 min). At t = 20 min, internal solution was changed to 3 m*M* Mg_i^{2+} and I_{Ca} declined. *Middle panel,* Current–voltage relationship for I_{Ca} obtained in the presence of 5 μ*M* cAMP and either 0.3 m*M* or 3 m*M* Mg_i^{2+}. *Bottom panel,* percentage inhibition of I_{Ca} by increasing $[Mg^{2+}]_i$ from 0.3 m*M* to 3 m*M* was calculated as a function of membrane potential

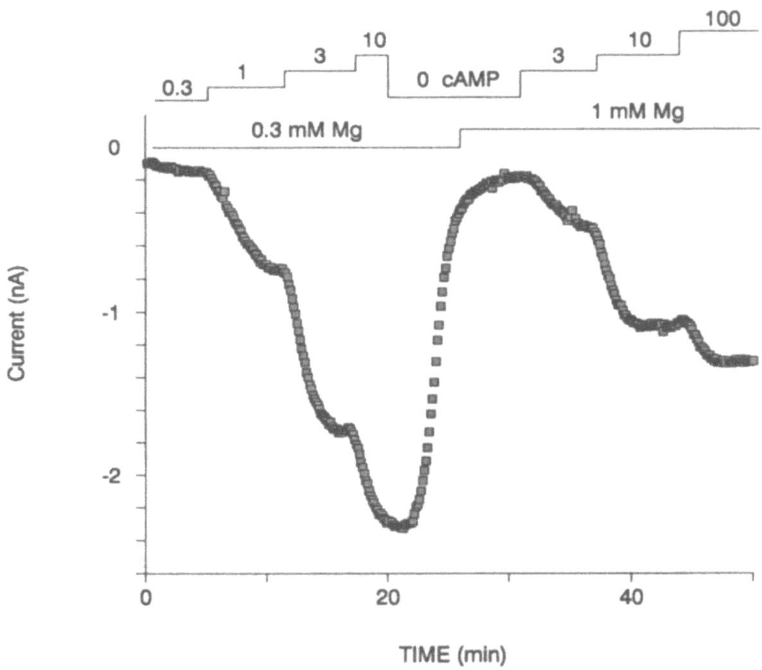

Fig. 11. Effect of $[Mg^{2+}]_i$ on the cAMP concentration-I_{Ca} response curve. Cell was perfused internally with different concentrations of cAMP as shown at the top of the figure and either 0.3 m*M* or 1 m*M* Mg_i^{2+}. Amplitude of I_{Ca} in presence of saturating concentrations of cAMP was reduced in the higher $[Mg^{2+}]_i$ compared with the lower $[Mg^{2+}]_i$

pathway. These experiments demonstrate that the effects of $[Mg^{2+}]_i$ are not on cAMP degradation or on the rate of channel phosphorylation.

Several experiments demonstrated that the effect of $[Mg^{2+}]_i$ was unlikely to be mediated by cAMP degradation. First, the effect of $[Mg^{2+}]_i$ on the cAMP concentration-I_{Ca} response curve was to decrease the maximum I_{Ca} that could be elicited, rather than to shift the concentration-response curve along the concentration axis, as would be expected if $[Mg^{2+}]_i$ increased the rate of cAMP degradation (Hartzell and Fischmeister 1986). This is shown in Fig. 11. I_{Ca} was measured at 0 mV, and the effect of different concentrations of cAMP was measured with different $[Mg^{2+}]_i$. In the presence of 0.3 m*M* Mg_i^{2+}, a maximal I_{Ca} of 2.3 nA was measured with concentrations of cAMP between 3 and 10 μ*M*. In contrast, with 3 m*M* Mg_i^{2+}, even with 100 μ*M* cAMP, the maximal I_{Ca} that could be elicited was 1.4 nA. The average effect of $[Mg^{2+}]_i$ on the cAMP concentration-response curve is shown in Fig. 12. The effect of increasing $[Mg^{2+}]_i$ is to decrease the maximal I_{Ca} with little change in the EC_{50} for cAMP. These results are inconsistent with a mechanism involving stimulation of phosphodiesterase activity by internal Mg^{2+}. Additional support for this conclusion is given by the finding that the effect of $[Mg^{2+}]_i$ on I_{Ca} elevated by

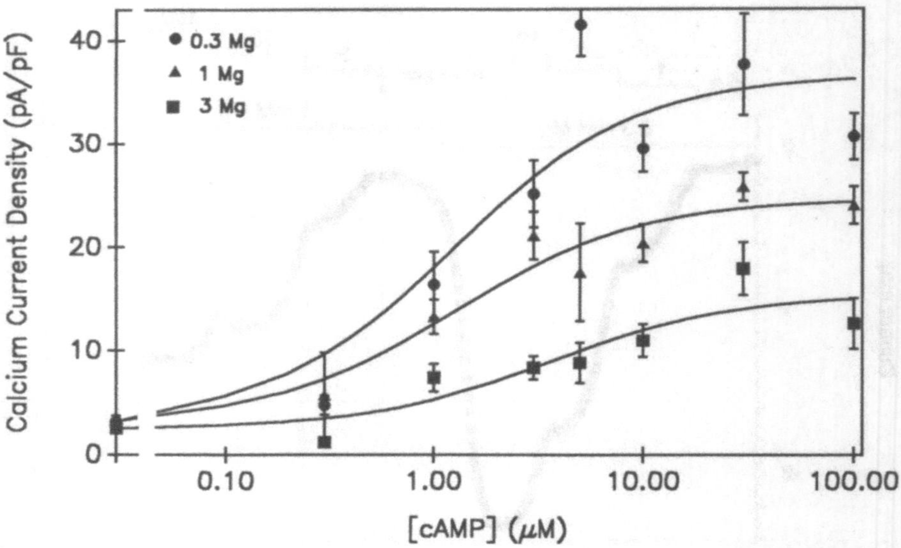

Fig. 12. cAMP concentration-I_{Ca} response curves obtained as shown in Fig. 11. *Circle*, 0.3 m*M* Mg_i^{2+}; *triangle*, 1 m*M* Mg_i^{2+}; *square*, 3 m*M* Mg_i^{2+}. I_{Ca} density was calculated as the net Ca^{2+} current divided by cell capacity. Each point is the mean of 3–48 cells, error bars are SEM. Reproduced from *Science* 239:778–780 (1988), copyright 1988 AAAS (White and Hartzell 1988)

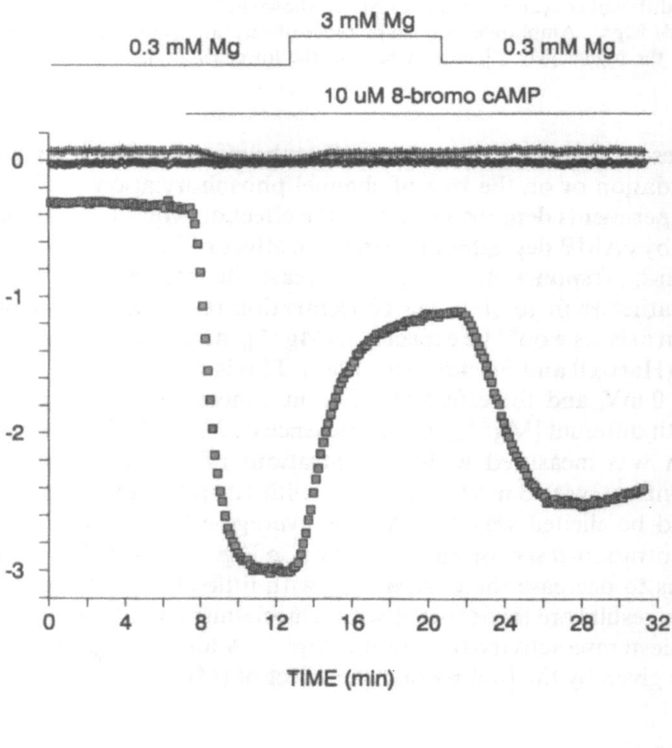

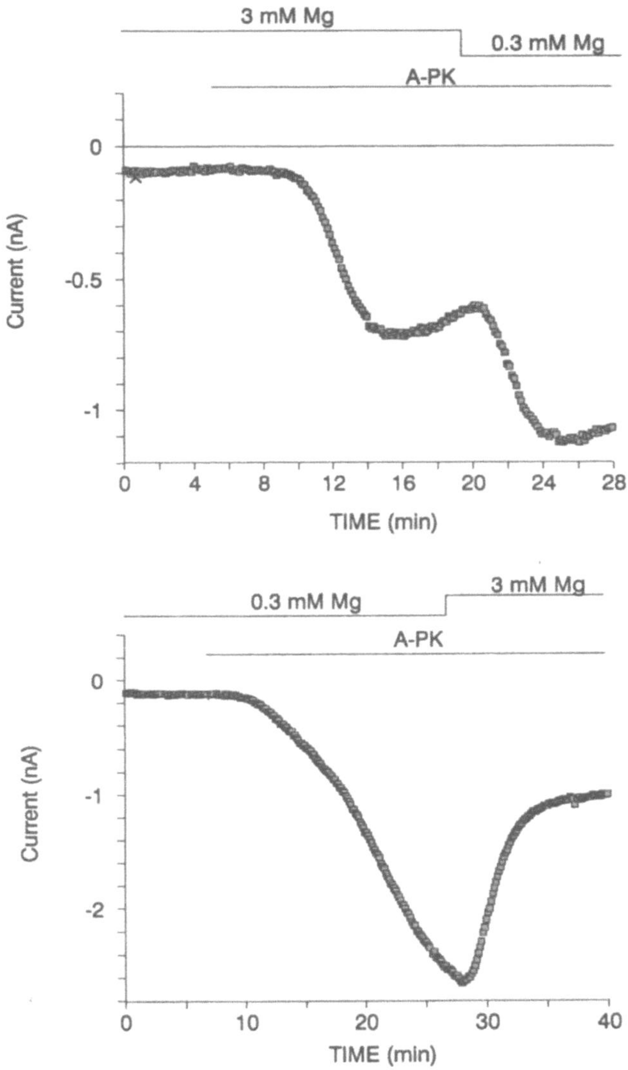

Fig. 14. Effect of changing $[Mg^{2+}]_i$ on I_{Ca} elevated by internal perfusion with catalytic subunit of cAMP-dependent protein kinase (provided by Dr. David Glass, Emory University). *Top panel,* $[Mg^{2+}]_i$ was increased from 0.3 m*M* to 3 m*M* after perfusion with cAMP-dependent protein kinase (75 μg/ml) and I_{Ca} increased. *Lower panel,* $[Mg^{2+}]_i$ was decreased and I_{Ca} decreased

◄

Fig. 13. Effect of changing $[Mg^{2+}]_i$ on I_{Ca} elevated by internal perfusion with 8-bromo-cAMP. At t=7 min, 10 μ*M* 8-bromo-cAMP was added to the internal perfusion. Increasing $[Mg^{2+}]_i$ from 0.3 m*M* to 3 m*M* during the period indicated produced a reversible decrease in I_{Ca}

internal perfusion with the stable analogue of cAMP, 8-bromo-cAMP, was virtually the same as that observed when I_{Ca} was elevated by native cAMP (Fig. 13). Increasing $[Mg^{2+}]_i$ fom 0.3 m*M* to 3.0 m*M* decreased cAMP-elevated I_{Ca} 63% ($n=8$) and 8-bromo-cAMP-elevated I_{Ca} 60% ($n=3$). Furthermore, $[Mg^{2+}]_i$ had a very similar effect on I_{Ca} that was elevated by internal perfusion with the catalytic subunit of cAMP-dependent protein kinase (Fig. 14). Decreasing $[Mg^{2+}]_i$ fom 3 m*M* to 0.3 m*M* (top panel) increased I_{Ca} elevated by the catalytic subunit of cAMP-dependent protein kinase. Increasing $[Mg^{2+}]_i$ from 0.3 m*M* to 3.0 m*M* (bottom panel) had the opposite effect. The effect of $[Mg^{2+}]_i$ on the amplitude of I_{Ca} elevated by cAMP-dependent mechanisms was easily reversible and occurred equally regardless of whether the change was from low to high Mg^{2+} or from high to low Mg^{2+}. In this regard, the effect of Mg^{2+} on the amplitude of I_{Ca} elevated by cAMP-dependent mechanisms differed from the effect of Mg^{2+} on the inactivation of I_{Ba}. As noted above, the effect of high $[Mg^{2+}]_i$ on I_{Ba} inactivation was usually not completely reversible.

Finally, the effect of $[Mg^{2+}]_i$ on I_{Ca} amplitude is not due to changes in protein phosphatase activity, because changing $[Mg^{2+}]_i$ produces identical results in the presence of okadaic acid, a potent protein phosphatase inhibitor (Hartzell, in preparation). In addition to decreasing the amplitude of cAMP-elevated I_{Ca}, $[Mg^{2+}]_i$ also altered the 200-ms inactivation curve for I_{Ca} (Fig. 15). The effect of $[Mg^{2+}]_i$ on the cAMP-elevated I_{Ca} inactivation curve was very similar to its effect on the basal I_{Ca} inactivation curve. The voltage at

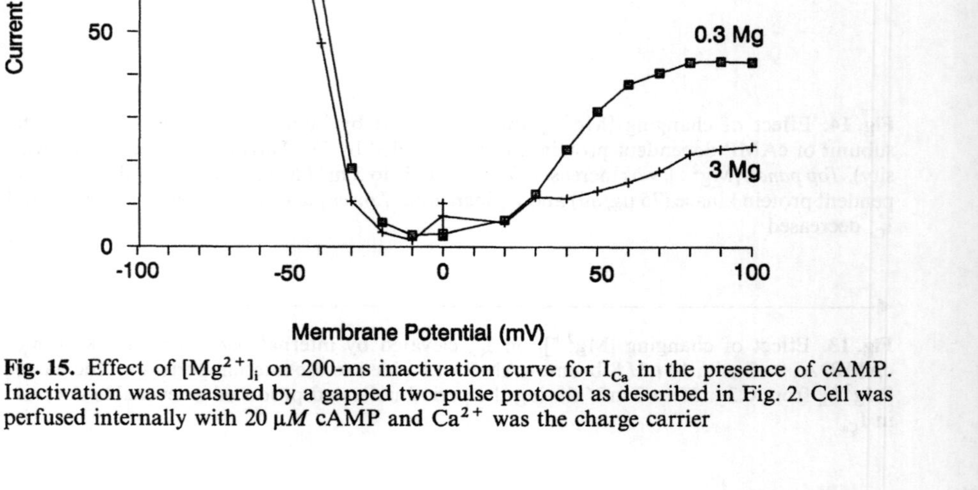

Fig. 15. Effect of $[Mg^{2+}]_i$ on 200-ms inactivation curve for I_{Ca} in the presence of cAMP. Inactivation was measured by a gapped two-pulse protocol as described in Fig. 2. Cell was perfused internally with 20 μ*M* cAMP and Ca^{2+} was the charge carrier

which 50% inactivation was observed was shifted slightly to negative potentials, and the relief from inactivation was strongly depressed with increasing $[Mg^{2+}]_i$.

Discussion

I_{Ca} is partly responsible for the plateau of the cardiac action potential, may contribute to the pacemaker depolarization in the sinoatrial node, and is responsible for the influx of Ca^{2+} ions that initiates excitation-contraction coupling in cardiac muscle (Hartzell 1989). I_{Ca} is modulated by cAMP-dependent protein phosphorylation by neurotransmitters and drugs that influence β-adrenergic and cholinergic receptors on cardiac cells (Tsien 1977; Reuter 1983; Tsien et al. 1986; Trautwein and Pelzer 1985, Fischmeister and Hartzell 1986; Hartzell and Fischmeister 1987; Hartzell 1988).

The several effects of $[Mg^{2+}]_i$ on current through Ca^{2+} channels differ in some respects but may share common mechanisms. First, $[Mg^{2+}]_i$ reduces the amplitude of I_{Ca}, I_{Ba}, and $I_{Ca(Na)}$ in a voltage-dependent manner. The inward current is decreased more strongly with more positive depolarizations, as would be expected if Mg^{2+} blocked the Ca^{2+} channel at a point whithin the voltage gradient of the membrane. The observation that the ability of internal Mg^{2+} to block inward current through Ca^{2+} channels is related to the charge carrier supports this mechanism. If Mg^{2+} competes for permeant ion binding to permeation sites in the channel in a voltage-dependent manner, one would predict that the effect of internal Mg^{2+} on current amplitude would be related to the relative affinities of permeant ions for permeation sites, which is $Ca^{2+} > Ba^{2+} > Na^{+}$ (Hille 1984). Permeant ions entering the cell would tend to compete for the Mg^{2+} ions at the inner mouth of the channel. The kinetics of this block is rather fast, because the effect is seen within 5 ms, which is the time-to-peak of the Ca^{2+} current. One expects a flickering blockade of single Ca^{2+} channels to be produced by internal Mg^{2+}. The finding that internal Mg^{2+} is much more effective at blocking I_{Ca} in the presence of cAMP than it is under basal conditions suggests that phosphorylation of the channel increases the affinity of the binding site for Mg^{2+}. Consistent with this suggestion is the observation that recovery from inactivation is decreased by cAMP in a manner similar to the decrease produced by increasing Mg^{2+}. One wonders whether the added negative charge on the channel introduced by phosphorylation might directly impose an additional attractive force on Mg^{2+} or whether the effect of phosphorylation is indirect via a conformational change. In any event, an increase in the affinity of the channel for Mg^{2+} is also indicated by preliminary experiments showing that when Na^{+} was the charge carrier in the presence of cAMP, Mg^{2+} decreased inward current amplitude drastically. For example, in the presence of 3 m*M* Mg_i^{2+}, cAMP often produced only negligible increases in $I_{Ca(Na)}$, in contrast to significant increases with 0.3 m*M* $[Mg^{2+}]_i$.

These effects of $[Mg^{2+}]_i$ on cAMP-elevated I_{Ca} amplitude were not due to effects of Mg^{2+} on channel phosphorylation, because they also occur when I_{Ca} is elevated by 8-bromo-cAMP or the catalytic subunit of cAMP-dependent protein kinase. These results rule out an effect of internal Mg^{2+} on phosphodiesterase, adenylate cyclase, or protein kinase activities.

The second documented effect of internal Mg^{2+} is an alteration of the inactivation kinetics and the steady-state inactivation of I_{Ca} and I_{Ba}. Increasing $[Mg^{2+}]_i$ causes a dramatic increase in the voltage-dependent inactivation of I_{Ba}. I_{Ca} inactivation in heart appears to be mediated by two mechanisms, one voltage-dependent and the other Ca^{2+}-dependent (Kass and Sanguinetti 1984; Eckert and Chad 1984; Lee et al. 1985; Bechem and Pott 1985; Kass et al. 1986; Chad and Eckert 1986). Some laboratories, however, have proposed that I_{Ca} inactivation is due entirely to a Ca^{2+}-dependent mechanism at least in snail neurons and GH_3 cells (Tillotson 1979; Eckert and Tillotson 1981; Chad and Eckert 1986), but other laboratories find only voltage-dependent inactivation (for example, Cohen and Lederer 1987). Both mechanisms clearly exist, but their proportion may depend upon the experimental conditions and the tissue studied. The existence of a Ca^{2+}-dependent mechanism is supported by the observations that inactivation increases with increasing intracellular or extracellular Ca^{2+} concentration, is greatly slowed when ions other than Ca^{2+} (such as Ba^{2+}) are used as the charge carrier, is reduced when intracellular Ca^{2+} is chelated, and is slowed when the driving force for Ca^{2+} entry is reduced (Eckert and Chad, 1984; Kass et al. 1986). The existence of "pure" voltage-dependent inactivation has been demonstrated by a two-pulse protocol in which the prepulse is given in the absence of Ca^{2+} (Akaike et al. 1988) or by using Na^+ as the charge carrier (Hadley and Hume 1987). The kinetics of I_{Ca} inactivation suggests that the rapid phase of inactivation is due to the Ca^{2+}-dependent component, whereas the slow phase is voltage dependent.

The U-shape of the inactivation curve (Fig. 2) can be viewed as the sum of voltage-dependent inactivation, which can be described by a Boltzman relation with incomplete inactivation, and Ca^{2+}-dependent inactivation, which has the same shape as the I_{Ca} I–V relationship. For short duration prepulses Ca^{2+}-dependent inactivation predominates, whereas for longer duration prepulses voltage-dependent inactivation predominates. It may be for this reason that some investigators (for example, Cohen and Lederer, 1987) fail to find clear evidence of Ca^{2+}-dependent inactivation under conditions of steady-state inactivation.

With either Ca^{2+} or Ba^{2+} as the charge carrier, increasing $[Mg^{2+}]_i$ shifts the inactivation curve to the left and decreases relief from inactivation. The shift in the inactivation curve could be caused by the ability of Mg^{2+} to screen negative surface charges on the internal surface of the membrane (Frankenhaeuser and Hodgkin 1957; Kass and Krafte 1987). A change in external divalent ion concentrations produces a shift in the I–V relationship and inactivation curves that reflects divalent cations binding to and screening negatively charged groups on the cell surface (Frankenhaeuser and Hodgkin 1957; Kass and Krafte 1987). In heart cells, a 10-fold change in $[Ca^{2+}]_o$ shifts the inactivation curve approximately 30 mV, whereas a 10-fold change in $[Mg^{2+}]_o$ shifts the curve only about 10 mV because of the lower affinity of Mg^{2+} for the surface charges (Kass and Krafte 1987). The shift in the inactivation curve is similar in magnitude but opposite in direction for the same changes in $[Mg^{2+}]_o$ and $[Mg^{2+}]_i$. That is, a 10-fold increase in $[Mg^{2+}]_i$ shifts the I–V relationship and the inactivation curves about 10 mV to the left, whereas a 10-fold increases in $[Mg^{2+}]_o$ shifts them 10 mV to the right. If this is true, then the internal surface charge is similar in magnitude to external surface charge. This might be expected considering the asymmetrical distribution of phospholipids in the plasma membrane: negatively charged phosphatidyl serines are often concentrated in the internal leaflet of the lipid bilayer. In addition, a negative internal surface charge may be partly due to the phosphate groups on phosphorylation sites on cytoplasmic domains of membrane proteins.

Although surface charge effects might explain the shifts in the inactivation and I–V curves, they are unlikely to explain the changes in relief from inactivation. Relief from inactivation observed with Ca^{2+} as the charge carrier is largely an expression of the fact that voltage-dependent inactivation is incomplete at voltages positive to 0 mV with low $[Mg^{2+}]_i$. The incomplete nature of voltage-dependent inactivation is demonstrated by the plateau of the steady-state inactivation curves with Ba^{2+} as the charge carrier and by the incomplete inactivation of I_{Ba} at most potentials. The incomplete inactivation of I_{Ba} with low $[Mg^{2+}]_i$ is not simply due to a shift in the inactivation parameters as a function of voltage, as would be expected from changes in surface charge. Rather, $[Mg^{2+}]_i$ seems to affect the amount of voltage-dependent inactivation at all voltages. Relief from inactivation is decreased by cAMP (Argibay et al. 1988) in a manner similar to that produced by increasing $[Mg^{2+}]_i$. It is worth considering that this decrease in relief from inactivation produced by cAMP might be related to accumulation of internal Mg^{2+} due to rapid hydrolysis of ATP for protein phosphorylation.

Although the kinetics of inactivation of I_{Ba} is significantly accelerated by increasing $[Mg^{2+}]_i$, I_{Ca} inactivation is not as dramatically affected. Indeed, if increasing $[Mg^{2+}]_i$ has any effect on I_{Ca} kinetics, it is to slow inactivation. This may seem puzzling, considering the fact that internal Mg^{2+} accelerates I_{Ba} inactivation. To explain these results, the following hypothesis is proposed. A single Ca^{2+} channel can be inactivated in either a voltage-dependent or Ca^{2+}-dependent manner, but not both. The rate of Ca^{2+}-dependent inactivation is determined by permeant Ca^{2+} ions, whereas the mode of inactivation is determined by the Mg^{2+} concentration at the cytoplasmic face of the channel. When Mg^{2+} is bound, the channel has a higher probability of inactivating by a voltage-dependent mechanism, whereas when Mg^{2+} is absent, the Ca^{2+}-dependent mechanism is favored. In the absence of Ca^{2+} as the charge carrier, Ca^{2+}-dependent inactivation is greatly slowed. The rate of inactivation of the macroscopic current is then determined by the fraction of the channels exhibiting voltage-dependent inactivation, which is set by the cytosolic Mg^{2+} concentration. In support of this hypothesis, it can be seen that I_{Ca} inactivation (at 0 mV) is often fitted by the sum of two exponentials with time constants of

~13 ms and ~30 ms (Argibay et al. 1988). Replacement of Ca^{2+} with Ba^{2+} results in the 13-ms component of inactivation slowing to 240 ms (Argibay et al. 1988). We believe that the 30-ms component represents voltage-dependent inactivation and the 13-ms component, Ca^{2+}-dependent inactivation. Increasing $[Mg^{2+}]_i$ causes a slowing of I_{Ca} inactivation due to an increase in the fraction of channels that exhibit voltage-dependent inactivation (the 30-ms component) rather than the Ca^{2+}-dependent component. In contrast, I_{Ba} inactivation is speeded by increasing $[Mg^{2+}]_i$ because more channels shift to the voltage-dependent inactivation mode (30-ms component) from the Ca^{2+}-dependent mode (240-ms component due to the absence of Ca^{2+} as the charge carrier).

Since the relative contribution of the 30-ms voltage-dependent component increases with increasing $[Mg^{2+}]_i$, one could conclude that voltage-dependent inactivation is dependent upon $[Mg^{2+}]_i$. An obvious prediction of this hypothesis is that voltage-dependent inactivation should be eliminated when $[Mg^{2+}]_i$ is zero. Under these conditions, I_{Ba} continued to exhibit a 30-ms inactivating component. It may be difficult to buffer $[Mg^{2+}]_i$ near the membrane to these low concentrations, and other ions may be able to substitute partially for Mg^{2+}. It is postulated that voltage-dependent inactivation is stimulated by $[Mg^{2+}]_i$ and that a portion of voltage-dependent inactivation is Mg^{2+}-dependent. There are several studies that apparently argue against this hypothesis. For example, cardiac muscle L-type channels incorporated into bilayers exhibit voltage-dependent inactivation in the absence of internal and external Mg^{2+} (Rosenberg et al. 1988). Further experimentation will be required to resolve these questions.

It remains unclear whether the decrease in I_{Ba} and $I_{Ca(Na)}$ amplitude produced by increasing $[Mg^{2+}]_i$ under basal conditions could be related to the more rapid inactivation under these conditions or whether different mechanisms mediate the effects of internal Mg^{2+} on inactivation and amplitude. The decrease in I_{Ca} amplitude by $[Mg^{2+}]_i$ in the presence of cAMP is unlikely to be related to the inactivation per se. It is possible that the binding of Mg^{2+} to the site in the permeation pathway that is responsible for the decrease in current amplitude results in a conformational change in the inactivation gate such that the voltage-dependent mode of channel inactivation is favored.

It has been pointed out by Lee et al. (1985) that voltage-dependent inactivation of Ca^{2+} channels may be an important mechanism of Ca^{2+} channel inactivation during an action potential, because the Ca^{2+} transient lags considerably behind the Ca^{2+} current, and consequently, Ca^{2+}-dependent inactivation might be small under these conditions. If this is true, changes in $[Mg^{2+}]$ that occur physiologically might have significant effects on Ca^{2+} currents during the action potential. Although $[Mg^{2+}]_0$ alters action potential shape (Watanabe and Dreyfus 1972; Chesnais et al. 1975), it is not yet known what effects $[Mg^{2+}]_i$ has on the cardiac action potential. Additional studies are required to determine whether $[Mg^{2+}]_i$ is physiologically regulated in cardiac cells, and if so, what effects this has on cardiac physiology (White and Hartzell 1989).

References

Akaike N, Tsuda Y, Oyama Y (1988) Separation of current- and voltage-dependent inactivation of calcium current in frog sensory neuron. Neurosci Lett 84:46–50

Argibay JA, Fischmeister R, Hartzell HC (1988) Inactivation, reactivation and pacing dependence of calcium current in frog cardiocytes: correlation with current density. J Physiol (Lond) 401:201–226

Bean BP (1985) Two kinds of calcium channels in canine atrial cells. J Gen Physiol 86:1–30

Bechem M, Pott L (1985) Removal of Ca current inactivation in dialysed guinea-pig atrial cardioballs by Ca chelators. Pflugers Arch 404:10–20

Chad JE, Eckert R (1986) An enzymatic mechanism for calcium current inactivation in dialysed Helix neurones. J Physiol (Lond) 378:31–51

Chesnais JM, Coraboeuf E, Sauviat MP, Vassas JM (1975) Sensitivity to H, Li, and Mg ions of the slow inward sodium current in frog atrial fibres. J Mol Cell Cardiol 7:627–642

Cohen NM, Lederer WJ (1987) Calcium current in isolated neonatal rat ventricular myocytes. J Physiol (Lond) 391:169–191

Duchatelle-Gourdon I, Hartzell HC, Lagrutta AA (1989) Modulation of the delayed rectifier potassium current in frog cardiomyocytes by β-adrenergic agonists and magnesium. J Physiol (Lond) 415:251–274

Eckert R, Chad JE (1984) Inactivation of Ca channels. Prog Biophys Mol Biol 44:215–267

Eckert R, Tillotson DL (1981) Calcium-mediated inactivation of the calcium conductance in caesium-loaded giant neurones of Apylsia californica. J Physiol (Lond) 314:265–280

Fischmeister R, Hartzell HC (1986) Mechanism of action of acetylcholine on calcium current in single cells from frog ventricle. J Physiol (Lond) 376:183–202

Fischmeister R, Hartzell HC (1987) Cyclic guanosine 3′,5′-monophosphate regulates the calcium current in single cells from frog ventricle. J Physiol (Lond) 387:453–472

Frankenhaeuser B, Hodgkin AL (1957) The action of calcium on the electrical properties of squid axons. J Physiol (Lond) 137:218–244

Godt RE, Lindley BD (1982) Influence of temperature upon contractile activation and isometric force production in mechanically skinned muscle fibers of the frog. J Gen Physiol 80:279–297

Hadley RW, Hume JR (1987) An intrinsic potential-dependent inactivation mechanism associated with calcium channels in guinea-pig myocytes. J Physiol (Lond) 389:205–222

Hamill OP, Marty A, Neher E, Sakmann B, Sigworth FJ (1981) Improved patch-clamp techniques for high-resolution current recording from cells and cell-free membrane patches. Pflugers Arch 391:85–100

Hartzell HC (1989) Regulation of cardiac ion channels by catecholamines, acetylcholine, and second messenger systems. Prog Biophys Mol Biol 52:165–247

Hartzell HC, Fischmeister R (1986) Opposite effects of cyclic GMP and cyclic AMP on Ca^{2+} current in single heart cells. Nature 323:273–275

Hartzell HC, Fischmeister R (1987) Effect of forskolin and acetylcholine on calcium current in single isolated cardiac myocytes. Mol Pharmacol 32:639–645

Hartzell HC, White RE (1989) Effects of magnesium on inactivation of the voltage-gated calcium current in cardiac myocytes. J Gen Physiol 94:745–767

Hille B (1984) Ionic channels of excitable membranes. Sinauer, Sunderland, MA, USA

Hodgkin AL, Huxley AF (1952) A quantitative description of membrane current and its application to conduction and excitation in nerve. J Physiol (Lond) 117:500–544

Horie M, Irisawa H (1987) Rectification of muscarinic K^+ current by magnesium ion in guinea pig atrial cells. Am J Physiol 253:H210–H214

Horie M, Irisawa H (1989) Dual effects of intracellular magnesium on muscarinic potassium channel current in single guinea-pig atrial cells. J Physiol (Lond) 408:313–332

Horie M, Irisawa H, Noma A (1987) Voltage-dependent magnesium block of adenosine-triphosphate-sensitive potassium channel in guinea-pig ventricular cells. J Physiol (Lond) 387:251–272

Kass RS, Krafte DS (1987) Negative surface charge density near heart calcium channels. Relevance to block by dihydropyridines. J Gen Physiol 89:629–644

Kass RS, Sanguinetti MC (1984) Inactivation of calcium channel current in the calf cardiac purkinje fiber. Evidence for voltage- and calcium-mediated inactivation. J Gen Physiol 84:705–726

Kass RS, Sanguinetti MC, Krafte DS (1986) Inactivation and modulation of cardiac Ca channels. In: Nathan RD (ed) Cardiac muscle: the regulation of excitation and contraction. Academic, New York, pp 29–53

Lee KS, Marban E, Tsien RW (1985) Inactivation of calcium channels in mammalian heart cells: joint dependence on membrane potential and intracellular calcium. J Physiol (Lond) 364:395–411

Matsuda H (1988) Open-state substructure of inwardly rectifying potassium channels revealed by magnesium block in guinea-pig heart cells. J Physiol (Lond) 397:237–258

Matsuda H, Saigusa A, Irisawa H (1987) Ohmic conductance through the inwardly rectifying K channel and blocking by internal Mg^{2+}. Nature 325:156–159

Mentrard D, Vassort D, Fischmeister R (1984) Calcium-mediated inactivation of the calcium conductance in cesium-loaded frog heart cells. J Gen Physiol 83:105–131

Pusch M, Conti F, Stühmer W (1989) Intracellular magnesium blocks sodium outward currents in a voltage- and dose-dependent manner. Biophys J 55:1267–1271

Reuter H (1983) Calcium channel modulation by neurotransmitters, enzymes and drugs. Nature 301:569–574

Rosenberg RL, Hess P, Tsien RW (1988) Cardiac calcium channels in planar lipid bilayers. L type channels and calcium-permeable channels open at negative membrane potentials. J Gen Physiol 92:27–54

Tillotson D (1979) Inactivation of Ca conductance dependent on entry of Ca ions in molluscan neurons. Proc Natl Acad Sci USA 76:1497–1500

Trautwein W, Pelzer D (1985) Gating of single calcium channels in the membrane of enzymatically isolated ventricular myocytes from adult mammalian hearts. In: Zipes DP, Jalife J (eds) Cardiac electrophysiology and arrhythmias. Grune and Stratton, Orlando, pp 31–42

Tsien RW (1977) Cyclic AMP and contractile activity in heart. Adv Cyclic Nucleotide Res 8:336–340

Tsien RW, Bean BP, Hess P, Lansman JB, Nilius B, Nowycky MC (1986) Mechanisms of calcium channel modulation by b-adrenergic agents and dihydropyridine calcium agonists. J Mol Cell Cardiol 18:691–710

Vandenberg CA (1987) Inward rectification of a potassium channel in cardiac ventricular cells depends on internal magnesium ions. Proc Natl Acad Sci (USA) 84:2560–2564

Watanabe Y, Dreyfus LS (1972) Electrophysiological effects of magnesium and its interactions with potassium. Cardiovasc Res 6:79–88

White RE, Hartzell HC (1988) Effects of intracellular free magnesium on calcium current in isolated cardiac myocytes. Science 239:778–780

White RE, Hartzell HC (1989) Magnesium ions in cardiac functions: regulator of ion channels and second messengers. Biochem Pharmacol 38:859–867

Blockage of Neuronal Low-Threshold Ca^{2+} Channels by Extracellular Mg^{2+} [1]

H. D. Lux [2] and E. Carbone [3]

Mg^{2+} belongs to the category of abundant extracellular and intracellular cations and has biologically relevant chemical properties (for review see Flatman 1984). Mg^{2+} ions are small compared with other earth alkaline ions and have a large hydration shell. Water substitution in Mg^{2+} aquocomplexes by ligands is known to occur by 3–4 orders of magnitude slower than in Ca^{2+} aquocomplexes (Diebler et al. 1969). Specificity of the binding of a comparably small metal ion like Mg^{2+} demands that the ligand be more tightly bound than the water to be substituted. Thus, Mg^{2+} ions are expected to pass with difficulty through small, water-filled channels of the excitable membrane. On the other hand, if Mg^{2+} ions are to compete effectively with other cations for passage, a ligand is required that binds the Mg^{2+} ion sufficiently tightly. Channels that are specifically permeable to divalent cations appear to be interesting candidates for studying these assumptions.

The aim of this study was to elucidate the action of Mg^{2+} ions on the neuronal, low-voltage-activated (lva) Ca^{2+} channel (Carbone and Lux 1987a). Instead of directly addressing the possibly complicated interference of Mg^{2+} with Ca^{2+} transport, the Na^{+} current through lva channels in the absence of other divalent cations was taken as a test of the action of external Mg^{2+}. It appears that submillimolar concentrations of external Mg^{2+} ions efficiently block the nonspecific cation transport through this Ca^{2+} channel, albeit less potently than Ca^{2+} ions (Fukushima and Hagiwara 1985; Tsien et al. 1987). The blocking action was pronounced, with negative inside membrane potentials, and was largely relieved at strong positive voltages (Lux et al. 1989). The kinetics of block relief and block establishment can give direct information of how fast and far Mg^{2+} ions can enter the channel, thus impeding the passage of other permeant ions. Additional charge-screening effects of the external membrane surface by Mg^{2+} could be clearly separated from this more specific action. The low apparent K_d of the Mg^{2+} block and its voltage dependence suggest that Mg^{2+} ions may interfere with Ca^{2+} transport and thus regulate Ca^{2+}-dependent activities near resting potentials.

---

[1] This work was partially supported by NATO (grant no. 0576/87) and by LIRCA Synthelabo Spa
[2] Abteilung Neurophysiologie, Max-Planck-Institut für Psychiatrie, Am Klopferspitz 18a, W-8033 Planegg, FRG
[3] Dipartimento di Anatomia e Fisiologia Umana, Corso Raffaello 30, I-10125 Torino, Italy

Methods

Sensory neurons (DRG cells) from thoracic and lumbar regions of 10-day-old chick embryos were used, and currents through lva channels were measured by methods described in detail by Carbone and Lux (1987a, b).

To block Na^+ and high-threshold Ca^{2+} channels, all external solutions contained 3 μ*M* tetrodotoxin (TTX) and 5 μ*M* ω-Conotoxin (ω-CgTX). TTX-insensitive Na^+ currents and high-voltage-activated ω-CgTx-insensitive Ca^{2+} currents could not be identified in DRG cells (McCleskey et al. 1987; Carbone and Lux 1988). Blockage of K^+ channels were ensured using K^+-free Ringers and intracellular solutions containing Na^+ or occasionally *N*-methylglucamine (NMG) as the main monovalent cation with the addition of tetraethylammonium (TEA). For the free divalent ion concentrations in the solutions, the following practical stability constants apply for the Mg^{2+} buffers at pH 7.5: $1.20 \times 10^2/M$ for ethyleneglycol-bis(oxyethylene-nitrile)-tetra-acetic acid (EGTA), $4.03 \times 10^4/M$ for *N*-(2-hydroxyethyl)-ethylenediamin,*N*,*N*,*N*-triacetic acid (HEDTA), and $1.51 \times 10^4/M$ for nitrilotriacetic acid (NTA). Usually, the same intra-(pipette) and extra-cellular Na^+ concentration (120 m*M*) was used to study inward as well as outward Na^+ currents through the lva Ca^{2+} channel.

Mg^{2+} Blockage of Ca^{2+} Currents Carried by Divalent Ions: A Survey

Mg^{2+} ions are not observed to carry significant currents through voltage-gated Ca^{2+} channels of different preparations (for review see Byerly and Hagiwara 1981; Tsien et al. 1987). The observation that Mg^{2+} ions hardly pass through the channel does not exclude the possibility that Mg^{2+} ions enter the channel to eventually block currents carried by other ions. Figure 1 shows that Ca^{2+} currents through the neuronal lva Ca^{2+} channel are indeed reduced when 1 m*M* Mg^{2+} is added to the bathing solution. The reduction is most pronounced at strongly negative membrane potentials. This is manifested by the reduced and shortened deactivation (tail) current on return to the holding potential observed after the depolarizing pulses had activated the Ca^{2+} currents. The concentrations of divalent ions are chosen here to approximate those in physiological media of vertebrate preparations.

Studies of the interference of Mg^{2+} ions with currents carried by divalent ions usually employed higher concentrations of blocking or transported ion species for different reasons. A comparison of the results could be ambiguous without considering the ionic conditions of the experiments. In addition, an extrapolation towards the behavior of the Ca^{2+} channels of a certain preparation from results received in another may be hampered by possibly different susceptibilities of the Ca^{2+} channel types to the Mg^{2+} block. With this precaution, a short survey of previous results on the Mg^{2+} block of Ca^{2+} channels will be given before returning to the situation of the lva Ca^{2+} channel.

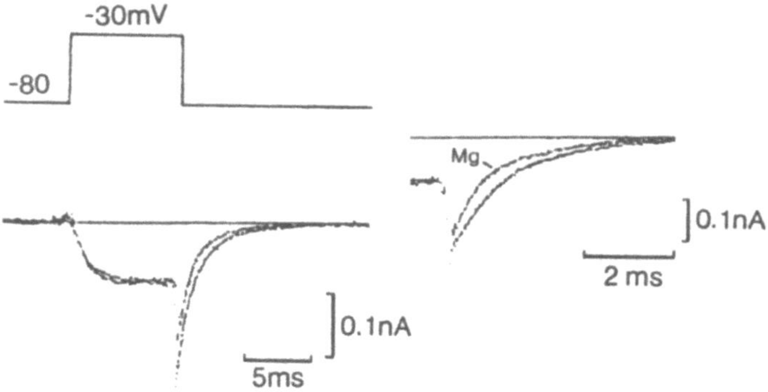

Fig. 1. Suppression by $[Mg^{2+}]_o$ of Ca^{2+} tail currents through low-voltage-activated (lva) channels of a chick dorsal root ganglion (DRG) cell. Superimposed currents during pulses to 0 mV in the presence and absence of Mg^{2+} with tail currents on repolarization to the holding potential of −80 mV. To the left, the tail currents at an expanded time scale. The bath contained 2 m*M* $[Ca^{2+}]_o$ with 120 m*M* Choline-Cl, the internal solution contained 120 m*M* CsCl. Note the shortening and reduction of the deactivation current when 1 m*M* $[Mg^{2+}]_o$ was added

An inhibitory action on inward Ca^{2+} currents was concluded early from $[Mg^{2+}]_o$-induced changes of the maximum rise of the calcium action potential in barnacle muscle fibers (Hagiwara and Takahashi 1967). In the presence of 100 m*M* Mg^{2+}, the apparent K_d of this $[Ca^{2+}]_o$-dependent measure was found to be between 25 and 40 m*M* $[Ca^{2+}]_o$, but was calculated to be below 15 m*M* Ca^{2+} in the absence of $[Mg^{2+}]_o$. The strength of the block was 1:46 of that of Co^{2+} ions, that are potent Ca^{2+} current blocker at and below millimolar concentrations. These results pointed to a rather weak Mg^{2+}-induced block of Ca^{2+} currents.

Ca^{2+} inward currents in snail neurons, measured at +25 mV membrane potential were found to be the same with or without 15 m*M* Mg_o^{2+} (Akaike et al. 1978). These experiments employed Ca^{2+} concentrations beyond the millimolar range. The relatively strong depolarization used in the described tests could have offset the action of Mg^{2+} ions, if the block was voltage-dependent in the manner suggested by the observation on the lva Ca^{2+} channel (Fig. 1).

Rates of entry of Mg^{2+} ions into cardiac Ca^{2+} channels have recently been derived from the analysis of unitary channel activity (Lansman et al. 1986). To optimally visualize the single channel events, relatively high Ba^{2+} concentrations (20–110 m*M*) were used in this study. Additional prolongation of mean open times of the channel by an agonist dihydropyridine (BAY K 8644) facilitated the analysis of the Mg^{2+}-induced effects. The blockage was manifested by closures, which reduced the mean open times. The shutting rate, measured as the reciprocal of the mean open time and evaluated as a function of $[Mg^{2+}]_o$, showed a slope of $1.9 \times 10^5\ M^{-1}\ s^{-1}$. This value likely represents a lower limit. It is highly probable that the pore could contain Ba^{2+} ions, thus rendering it frequently inaccessible for the blocking Mg^{2+} ion in a single file process. Extrapolating to low $[Ba^{2+}]_o$ under the assumption of a single binding site would probably increase this estimate to nearly 1×10^6 M^{-1} s^{-1}. The Mg^{2+} exit rates, which were taken as the inverse of the mean closed times (times to unblock of the channel), ranged between 1 and 3.5×10^3 s^{-1}. The ratio of the exit and entry values corresponds to K_d's of the Mg^{2+}-induced block in the millimolar range. The patch-to-patch comparison of the Mg^{2+}-induced block probably prohibited the resolution of eventual effects of $[Mg^{2+}]_o$ lower than 1 m*M*. It should be noted that the values were derived at a fixed membrane potential (0 mV). The block was not investigated for a possible voltage dependence.

Blockage of Currents Carried by Monovalent Ions Through Ca^{2+} Channels

In the absence of divalent cations, Ca^{2+} channels are known to carry monovalent cations with low specifity but high efficacy (Kostyuk and Krishtal 1977; for review see Tsien et al. 1987).

Unitary Li^+ currents through cardiac Ca^{2+} channels were found to be blocked by Mg_o^{2+} (Lansman et al. 1986), but to produce significant effects, 1 m*M* Mg^{2+} had to be applied at minimum. This complies with results of Almers et al. (1984) in nonspecific (Na^+) cationic currents through Ca^{2+} channels in skeletal muscle fibers. However, a relatively strong block by Mg^{2+} of Na^+ currents through cardiac Ca^{2+} channels is reported by Matsuda (1986). With a K_d of 60 μ*M* at −33 mV, the block by Mg^{2+} was still 1/50 of that by Ca^{2+}. In the lva Ca^{2+} channel of chick sensory neurons (Lux et al. 1989) as well as of neoplastic lymphocytes (Fukushima and Hagiwara 1985), the K_d of the Mg^{2+}-induced block appears somewhat smaller than that found in heart muscle.

The lva channel becomes well activated at relatively negative membrane potentials, e.g., at −45 mV at in Fig. 2 (see also Fig. 3A). In studies of unitary

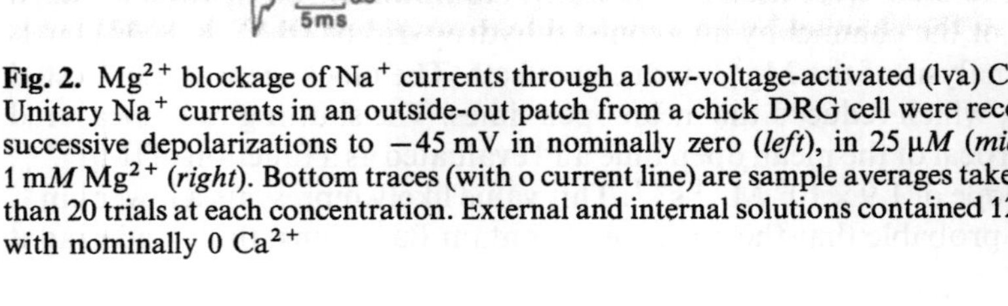

Fig. 2. Mg^{2+} blockage of Na^+ currents through a low-voltage-activated (lva) Ca^{2+} channel. Unitary Na^+ currents in an outside-out patch from a chick DRG cell were recorded during successive depolarizations to −45 mV in nominally zero (*left*), in 25 μ*M* (*middle*), and in 1 m*M* Mg^{2+} (*right*). Bottom traces (with o current line) are sample averages taken from more than 20 trials at each concentration. External and internal solutions contained 120 m*M* NaCl with nominally 0 Ca^{2+}

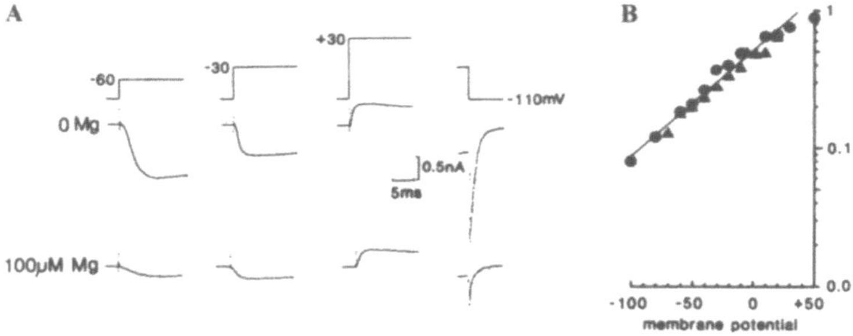

Fig. 3A, B. Blocking of whole cell low-voltage-activated (lva) Na^+ currents by Mg^{2+}. **A** Examples of currents through lva channels in the absence of divalent ions (*upper traces*) and with 100 μ*M* Mg^{2+} added (*lower traces*) to the bath solution, which contained 120 m*M* NaCl. 120 μ*M* NaCl was also applied internally. Note the strong decrease of currents at more negative membrane potentials, such as at −60 mV and of tail currents on repolarization to the holding potential of −110 mV (to the right). **B** Potential-dependent suppression of Na^+ currents by Mg^{2+}. The ratios of the current amplitudes at 100 μ*M* Mg_o^{2+} to that in divalent cation-free solution at the same membrane potential are plotted (*filled circles*) against membrane potential on semilogarithmic scale. 120 m*M* NaCl externally and internally. The *triangles* display data of Fukushima and Hagiwara (1985) showing the effect of 80 μ*M* $[Mg^{2+}]_o$ on Na^+ currents through Ca^{2+} channels in mouse neoplastic B lymphocytes, under comparable ionic conditions

Na^+ currents through the lva Ca^{2+} channel in isolated membrane patches, it was found that $[Mg^{2+}]_o$ as low as 25 μ*M* significantly reduces the channel activity at this membrane potential (Fig. 2, middle row) and 1 m*M* Mg^{2+} suppresses the activity to a low percentage of the control. The effect is manifested by prolonged channel closures if these are compared with spontaneous closures of the control in the absence of divalent ions. The unitary current amplitudes show little if any effect. The features are similar to those observed with Ca^{2+} ions as blockers (Carbone and Lux 1987b, c; Lux and Carbone 1987), which, however, show a K_d of the block of 2 μ*M* as compared to the K_d for Mg^{2+} of about 40 μ*M* at this potential.

The prolongation of mean closed times to 0.9 ms from a value of 0.65 ms in the control in the data at 25 μ*M* of Fig. 2 may be attributed to times during which the channel is blocked by Mg^{2+} ions. Under this assumption a rate of Mg^{2+} entry of 0.43 ms^{-1} with 25 μ*M* Mg_o^{2+} or a molar rate of 17×10^6 M^{-1} s^{-1} is calculated. The simple approach tacitly assumes that only the open channel is subjected to block. This is refuted, however, by the observation that mean open times were not proportionally shortened, not even during the stronger block exerted by higher $[Mg^{2+}]_o$. The difficulty is removed by employing the suggestion that the channel can also be blocked if it was closed. With this assumption, the estimate of the molar entry rate is reduced to a value 6×10^6 M^{-1} s^{-1}, which was the mean received from observations at 25 and 100 μ*M* Mg^{2+}. This value is by at least an order of magnitude larger than that deduced from observations on the cardiac Ca^{2+} channel. Blocking rate and K_d of the block (see Fig. 3) would result in a Mg^{2+} exit rate of 3×10^2 s^{-1} for the lva channel at −45 mV.

The discrepancy in entry rates of lva and cardiac Ca^{2+} channels may in part arise from the different conditions of the tests. With relatively high blocker concentration, the block could approach saturation. If not every blocker ion finds access to the channel an underestimation of the blocking or entry rates would result. However, a comparison of the data where this is possible (~1 m*M* Mg^{2+}) suggests that the block is somewhat stronger in the case of the neuronal lva channel. This conclusion remains if a potential dependent K_d is assumed to hold for the data on the cardiac Ca^{2+} channel as it is suggested by the observations on the lva Ca^{2+} channel.

Voltage Dependence of Mg^{2+} Blockage

A strong voltage dependence of the block by Mg^{2+} of Na^+ currents through Ca^{2+} channels was observed by Fukushima and Hagiwara (1985) in whole cell recordings of neoplastic lymphocytes (Fig. 3 B). Also in unitary Na^+ currents through cardiac Ca^{2+} channels, the block was found to be more severe at negative potentials (−40 mV) than at slightly positive potentials (Lansman et al. 1986). In whole neuron recordings of lva Na^+ currents, the block by external Mg^{2+} was entirely similar to that found in lymphocytes. It continued to increase when the membrane potential was made more negative (Fig. 3) and was seen clearly also in the deactivation (tail) currents. The block was measured as the fraction of unblocked channels with controls made in the absence of divalent ions (Fig. 3 B). The slope of the block suggests an exponential voltage dependence with an e-fold increase over a change of −50 mV.

This voltage dependence is different for Mg^{2+} than for other Ca^{2+} channel blockers, including metal ions (Lansman et al. 1986) and the Ca^{2+} ion itself. Except for Mg^{2+} membrane hyperpolarization is known to lessen the steady state block of Na^+ currents through Ca^{2+} channels. Thus, at strongly negative membrane potentials (−100 mV), even low Mg^{2+} concentrations can block as potently as equimolar Ca^{2+} (Lux et al. 1989).

It should be noted that raising $[Mg^{2+}]_o$ from 0 to 100 μ*M* produces little change of the activation-voltage dependence. At 1 m*M* $[Mg^{2+}]_o$, the shift in the I/V relationship of current activation was about +10 mV. The observations speak against a principal explanation of the block by a voltage shift of current activation due to screening by Mg^{2+} of surface charges of the membrane.

Discussion

The results from neuronal lva Ca^{2+} channel studies show that external Mg^{2+} concentrations produce two effects. One consists of some voltage shift of the activation time course of the voltage-gated channel as expected from the reduction of the external surface potential by a screening of negative charges.

This nonspecific effect is similar but smaller than that observed with $[Ca^{2+}]_o$ and Na^+ channels in squid axons (see Frankenhäuser and Hodgkin 1957) and on Ca^{2+} channels in *Helix* neurons (Kostyuk et al. 1982).

The strong voltage dependence of the blocking action of $[Mg^{2+}]_o$ and nonspecific cationic (Na^+) currents through the lva Ca^{2+} channel points to a separate mechanism which senses the electric field and is thus located inside the channel. It is apparent that Mg^{2+} ions have to enter the channel to reach the field-sensitive site, although they can hardly pass through the full extent of the channel. This latter conclusion is necessary to comply with the observation of an insignificant Mg^{2+} transport through the channel.

It is surprising, however, that Mg^{2+} ions enter the channel with remarkable ease, although less well than other divalent cations such as Ca^{2+} and Sr^{2+}, and then Na^+ ions. This is not expected from their chemical nature, as the considerable activation energy necessary for water substitution differs by three orders of magnitude from that of other earth alkaline ions including Ca^{2+} ions. It thus appears that only part of the hydration shell needs to be removed for Mg^{2+} entry. The remaining Mg^{2+}-water complex may be still prevent internal Mg^{2+} exit and thus Mg^{2+} transport through the Ca^{2+} channel. Mg^{2+} blocks the nonspecific cationic passage through the channel most probably because of its slow exit from the binding site (towards the external surface). This implies that the Mg^{2+} ion binds with considerably specifity to the site within the channel. Its blocking action may, in fact, compare with that of the natural ion of the Ca^{2+} channel, suggesting that Mg^{2+} ions could also interfere with the Ca^{2+} transport. This holds especially under the condition of negative membrane potentials at which lva Ca^{2+} channels are known to be significantly activated. Since Mg^{2+} ions can leave the channel only against the electric field in this situation, a long-lasting sojourn of Mg^{2+} ions inside the channel and a potent competition with Ca^{2+} entry should be expected. This voltage-dependent rectifying action of Mg^{2+} on Ca^{2+} currents may, in fact, be significant at resting potentials regarding the known inhibitory effects of Mg^{2+} on spontaneous, Ca^{2+}-dependent synaptic activity.

References

Akaike N, Lee KS, Brown AM (1978) The calcium currents of *Helix* neuron. J Gen Physiol 71:509–532

Almers W, McCleskey EW, Palade PT (1984) A non-selective cation conductance in frog muscle membrane blocked by micromolar external calcium ions. J Physiol (Lond) 353:565–583

Byerly L, Hagiwara S (1981) Calcium channel. Ann Rev Neurosci 4:69–125

Carbone E, Lux HD (1987a) Kinetics and selectivity of a low voltage-activated calcium current in chick and rat sensory neurones. J Physiol (Lond) 386:547–570

Carbone E, Lux HD (1987b) Single low voltage-activated calcium channels in chick and rat sensory neurones. J Physiol (Lond) 386:571–601

Carbone E, Lux HD (1987c) External Ca^{2+} ions block unitary Na^+ currents through Ca^{2+} channels of cultured chick sensory neurones by favouring prolonged closures. J Physiol (Lond) 382:124P

Carbone E, Lux HD (1988) ω-Conotoxin blockade distinguishes Ca from Na permeable states in neuronal calcium channels. Pflügers Arch 413:14–22

Diebler H, Eigen M, Ilgenfritz G, Maas G, Winkler R (1969) Kinetics and mechanism of reactions of main group metal ions with biological carriers. Pure Appl Chem 20:93–115

Flatman PW (1984) Magnesium transport. J Membr Biol 80:1–14

Frankenhäuser B, Hodgkin AL (1957) The action of calcium on the electrical properties of squid axons. J Physiol (Lond) 137:218–244

Fukushima Y, Hagiwara S (1985) Currents carried by monovalent cations through calcium channels in mouse neoplastic B lymphocytes. J Physiol (Lond) 358:255–284

Hagiwara S, Takahashi K (1967) Surface density of calcium ions and calcium spikes in the barnacle muscle fiber membrane. J Gen Physiol 50:583–601

Kostyuk PG, Krishtal OA (1977) Effects of calcium and calcium-chelating agents on the inward and outward current in the membrane of mollusc neurones. J Physiol (Lond) 270:569–580

Kostyuk PG, Mironov SL, Doroshenko PA (1982) Energy profile of the calcium channel in the membrane of mollusc neurons. J Membr Biol 70:181–189

Lansman JB, Hess P, Tsien RW (1986) Blockade of Ca current through single calcium channels by Cd^{2+}, Mg^{2+}, and Ca^{2+}: Voltage and concentration dependence of calcium entry into the pore. J Gen Physiol (Paris) 88:321–347

Lux HD, Carbone E (1987) External Ca ions block Na-conducting Ca channel by promoting open to closed transitions. In: Ovchinnikov YA, Hucho E (eds) Receptors and ion channels. De Gruyter, New York, pp 149–155

Lux HD, Carbone E, Zucker H (1989) Block of Na^+ ion permeation and selectivity of Ca channels. Ann NY Acad Sci (USA) 560:94–102

Matsuda H (1986) Sodium conductance in calcium channels of guinea-pig ventricular cells induced by removal of external calcium ions. Pflügers Arch 407:465–475

McCleskey EW, Fox AP, Feldman DH, Cruz LJ, Olivera BM, Tsien RW, Yoshikami D (1987) ω-Conotoxin: direct and persistent blockade of specific types of calcium channels in neurons but not in muscle. Proc Natl Acad Sci USA 84:4327–4331

Tsien RW, Hess P, McCleskey EW, Rosenberg RL (1987) Calcium channels: mechanisms of selectivity, permeation, and block. Annu Rev Biophys Chem 16:265–290

Dual Block by Mg^{2+} of the *N*-Methyl-D-Aspartate-Activated Channel

John W. Johnson[1,2] and Philippe Ascher[1]

Fast excitatory neurotransmission appears to be mediated by a relatively small number of receptor types. Several of these are activated by the neurotransmitter candidate L-glutamate. Two groups of glutamate receptors mediating fast excitatory neurotransmission can be clearly distinguished. The best studied group is specifically activated by *N*-methyl-D-aspartate (NMDA). This receptor is highly voltage dependent; under physiological conditions, current flow is nearly fully blocked at a neuron's normal resting potential. The second group, which is probably more heterogeneous, exhibits very little voltage dependence and appears to mediate synaptic transmission under most circumstances at a majority of fast excitatory CNS synapses. These receptors are often simply called non-NMDA receptors. The difference between the voltage dependence of NMDA and non-NMDA responses is due to the block of NMDA-activated channels by Mg^{2+} ions.

Although NMDA receptors are generally unable to mediate significant excitatory responses at a cell's resting potential, their voltage dependence allows activation at depolarized potentials. Thus, the voltage dependence of NMDA receptors serves to reserve their use for situations in which the postsynaptic cell is depolarized, for example, during high frequency stimulation (Herron et al. 1986). A possible reason for this exclusiveness is suggested by the second unusual property of NMDA receptors: Their channel is highly permeable to Ca^{2+} (MacDermott et al. 1986). Because the NMDA receptor is both Ca^{2+} permeable and voltage dependent, transmission at glutamatergic synapses can result in a large Ca^{2+} influx, but only following strong excitation.

The physiological and pathological consequences of these properties of glutamatergic transmission are probably far-reaching. The Ca^{2+} influx through NMDA receptors that occurs during high frequency stimulation can result in long-term potentiation of synaptic transmission (LTP) at some synapses (see Bliss and Lynch 1988). This phenomenon is believed to be involved in some forms of memory. NMDA receptors are thought to be similarly involved in the selective stabilization of synapses during development (see, e.g., Cline et al. 1987). Excessive stimulation of Ca^{2+} influx through NMDA recep-

[1] Laboratoire de Neurobiologie, Ecole Normale Supérieure, 46, rue d'Ulm, 75230 Paris, cédex 05, France
[2] Present address: Department of Behavioral Neuroscience, 446 Crawford Hall, University of Pittsburgh, Pittsburgh, PA 15260, USA

tors on depolarized neurons may be the cause of cell death following a variety of CNS injuries (see Choi 1988). The mechanisms responsible for the voltage dependence of the NMDA response thus play an important role in the CNS.

The block of the NMDA-activated channel at hyperpolarized voltages is mediated by Mg^{2+} ions in the extracellular solution. In addition, Mg^{2+} ions in the cytoplasm cause a voltage-dependent block at depolarized voltages. This chapter will review these two actions of Mg^{2+} ions on the NMDA response.

Block by Extracellular Mg^{2+}

A depressant action of Mg^{2+} on NMDA-induced responses was demonstrated in 1980 by Ault et al. The voltage dependence of this blockage, however, was not described until the advent of patch clamp techniques (Hamill et al. 1981), which allowed high quality voltage clamp of vertebrate neurons and the direct observation of the current flowing through single ionic channels. At the whole cell level, the NMDA-induced conductance change was observed to be nearly voltage independent in the absence of extracellular Mg^{2+}. When extracellular Mg^{2+} in the low millimolar range was added, the response remained linear at voltages positive of 0 mV, but at negative voltages a block that increased with hyperpolarization was observed (Nowak et al. 1984; Mayer et al. 1984; Mayer and Westbrook 1987; Ascher and Nowak 1988). In single channel recordings (Nowak et al. 1984; Ascher and Nowak 1988), Mg^{2+} caused channel openings to be interrupted by brief closures (channel flicker; Fig. 1). The time the channel stays open between these brief closures decreased, while the duration of the closures increased, with hyperpolarization. In both whole cell and single channel experiments, the magnitude of the Mg^{2+} blocking action was observed to depend on the Mg^{2+} concentration.

Model of Mg^{2+} Block of Single Channel Openings

The model that was used to explain the blockage by Mg^{2+} started with the simple idea that when Mg^{2+} is bound to a site associated with the NMDA receptor, current flow is prevented. One way to account for the voltage dependence of the block is to hypothesize that the Mg^{2+} binding site is at a location where it can sense the voltage drop across the plasma membrane. Because the voltage drop between the intracellular and extracellular solutions takes place across the plasma membrane, there is a very strong electrical field within the membrane. This membrane field will influence the free energy of any charged particle and therefore also its dissociation constant from a binding site within the field. The most likely location for a blocking site within the membrane field would seem to be within the NMDA-activated channel.

This simple model predicts that the dissociation constant of Mg (K_{Mg}) from a binding site located within the membrane field will vary exponentially with

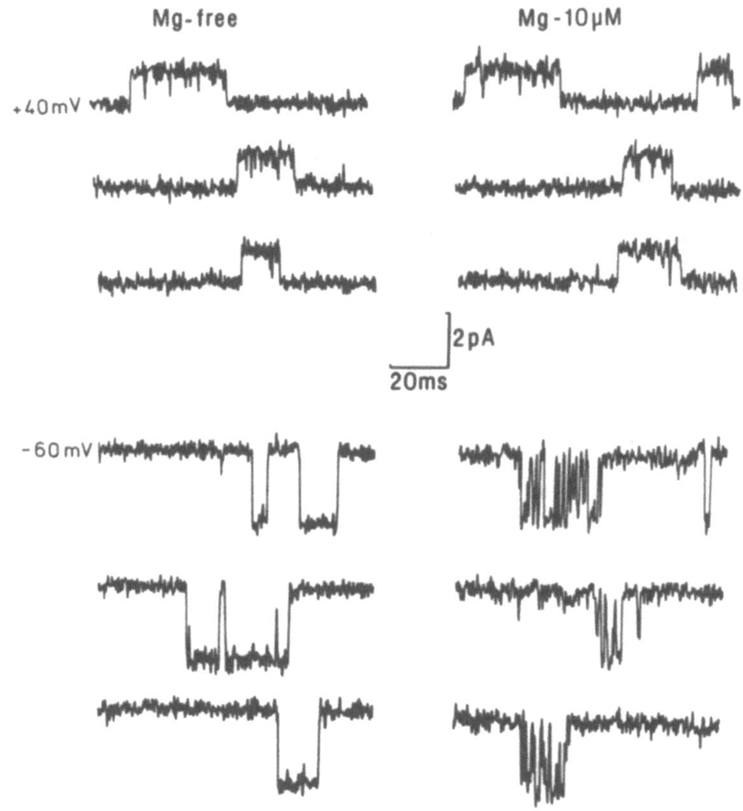

Fig. 1. Flicker of *N*-methyl-D-aspartate (NMDA)-activated channels is induced at hyperpolarized voltages by extracellular Mg^{2+}. This recording is made using an outside-out patch. *Upper half,* 10 μ*M* Mg^{2+} can be seen to have no significant effect on channel openings at +40 mV. *Lower half,* 10 μ*M* Mg^{2+} induces brief closures (flicker) in channel openings at −60 mV. In this and subsequent figures, recordings were made from neurons in primary culture taken from 15- to 16-day-old mouse embryos. Reprinted by permission from *Nature* 307: 463, 1984, Macmillan Magazines Ltd

membrane voltage:

$$K_{Mg} = K_0 \exp(V/V_o) \quad (1)$$

where K_o is the dissociation constant of Mg^{2+} from its binding site in the absence of a membrane voltage, V the transmembrane voltage, and V_o is a measure of the steepness of the voltage dependence of the block. V_o can be used as a parameter to describe the block empirically. Alternatively, it can be used to estimate the location of the binding site within the membrane field (Woodhull 1973) if one makes the following assumptions: 1) channel block is accomplished by only one blocking ion, 2) no other ions interfere with the block, and 3) the blocking ion leaves the blocking site by the same route that it entered (in this case to the extracellular solution). The model then predicts that K_{Mg} will vary exponentially with the fraction of membrane voltage felt at the binding site, or:

$$V_o = RT/\delta z F \tag{2}$$

where δ is the fraction of the membrane field felt at the binding site measured from the source of Mg^{2+} (in this case the extracellular surface of the membrane), z the valence of the blocking ion, F Faraday's constant, R the gas constant, and T the absolute temperature.

In order to apply this model to the channel flicker observed by Nowak et al. (1984), the rates at which Mg^{2+} ions bind to and unbind from the blocking site must be considered:

$$R \underset{\alpha}{\overset{\beta}{\rightleftharpoons}} R^* + Mg \underset{k_-}{\overset{k_+}{\rightleftharpoons}} Mg\,R \tag{3}$$

where R is the closed receptor with agonist(s) bound (the interaction between agonist and receptor is not considered here and would occur to the left of the above diagram), R* the open unblocked channel, β the rate constant for channel opening, α the rate constant for closing of the open and unblocked channel, Mg R the blocked channel, k_+ the rate constant for the binding of Mg^{2+} ions to the blocking site, and k_- the rate constant for unbinding. In this scheme, the apparent dissociation constant of Mg^{2+} is related to the binding rate constants by:

$$K_{Mg} = k_-/k_+ \tag{4}$$

The dependence of k_+ and k_- on the Mg^{2+} concentration and voltage can be predicted if the rate of Mg^{2+} ion entry into and exit from the channel is assumed to be limited by an energy barrier, between the extracellular solution and the binding site, over which the ions must pass (Fig. 2; Woodhull 1973). In the absence of applied voltage, the height of this barrier will determine k_+, while the energy difference between the barrier's peak and the binding site will determine k_-. When a voltage is applied, k_+ will also depend on the fractional membrane field felt at the peak of the barrier (δ_1, calculated from V_{o_1}), while k_- will depend on the fractional membrane field between the peak of the barrier and the binding site (δ_2, calculated from V_{o_2}):

$$k_+ = k_{+o} \exp(-V/V_{o_1}) \quad \text{and} \quad V_{o_1} = RT/\delta_1 z F \tag{5}$$

$$k_- = k_{-o} \exp(V/V_{o_2}) \quad \text{and} \quad V_{o_2} = RT/\delta_2 z F \tag{6}$$

where k_{+o} is the on rate and k_{-o} the off rate in the absence of applied voltage. As would be hoped, by combining Eqns. (4)–(6), one can reproduce Eqn. (1).

Ascher and Nowak (1988) were able to relate the experimentally observed channel flicker to the values of k_+ and k_- by using the same interpretation of channel flicker as did Neher and Steinbach (1978) in their study of the block of the acetylcholine (ACh)-activated channel by local anesthetics. During

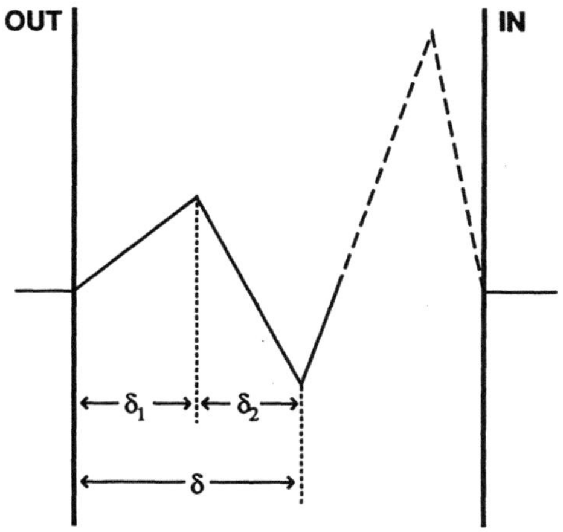

Fig. 2. Schematic energy profile showing a single binding site for a blocking ion within a channel. *Heavy vertical lines* show the outer (*left*) and inner (*right*) edges of the membrane. Height of the jagged line is proportional to the free energy of an ion in that position in the channel. Membrane potential is 0. An ion entering from the extracellular solution must travel a distance δ_1 to reach the peak of the energy barrier, then an additional distance δ_2 to reach the energy minimum (binding site). Distance from the outer edge of the membrane to the binding site ($\delta_1+\delta_2$) is equal to δ. When the ion leaves the binding site, it will usually leave to the outside because of the presence of a much higher barrier to the inside. Inner barrier is shown as a *dashed line* to indicate that other binding sites may exist for ions entering from the intracellular solution

channel flicker in the presence of a low concentration of Mg^{2+}, the channel appears to jump between the fully open and fully closed state (Fig. 1). The brief intervals during which the channel was fully closed were interpreted as periods of blockage by Mg^{2+}. The average length of these intervals should depend on the rate at which Mg^{2+} leaves its binding site. The brief intervals during which the channel was fully open were interpreted as periods when the channel was open, and Mg^{2+} was not occupying its blocking site. The average length of these intervals should depend on both the rate at which the channel closes and the rate at which Mg^{2+} enters its binding site. The mean blocked time (τ_c) and the mean unblocked, open time (τ_o) of the channel were measured directly from single channel records over a range of voltages and Mg^{2+} concentrations. They were then related to k_- and k_+ by:

$$k_- = (\tau_c)^{-1}$$

$$k_+ = ((\tau_o)^{-1} - \alpha)/[Mg^{2+}]_o$$

where α is, as in Eqn. (3), the rate of channel closing (measured in the absence of Mg^{2+}) and $[Mg^{2+}]_o$, the extracellular Mg^{2+} concentration. It was found that k_- is independent of $[Mg^{2+}]_o$, that k_+ varies linearly with $[Mg^{2+}]_o$, and that both k_- and k_+ vary exponentially with membrane voltage, as predicted by the model. From their measurements of the dependence of k_- and k_+ on

V, the following estimates can be made: V_{o_1}, 17 mV (δ_1, 0.76); V_{o_2}, 47 mV (δ_2, 0.27); V_o, 12.5 mV (δ, 1.0). The dissociation constant of Mg^{2+} at $V=0$ was estimated to be 8.8 mM.

The value $\delta=1$ places the Mg^{2+} binding site at the intracellular edge of the NMDA-activated channel. However, the very different values of δ_1 and δ_2, though not a priori impossible, suggest a barrier asymmetry that was not expected based on previous studies of channel blockers (e.g., Neher and Steinbach 1978). Ascher and Nowak (1988) therefore expressed doubt about the validity of this estimate of the binding site location. Another reason to doubt this estimate of δ will be presented later.

Along with the apparent asymmetry of the barrier energy profile, another clear difference between the block of the NMDA-activated channel by Mg^{2+} and that of the nicotinic ACh-channel by local anesthetics was observed. In local anesthetic block, when the channel spends more time blocked, it also spends a proportionally longer time flickering between the open, blocked state and open, unblocked state. This was interpreted (Neher and Steinbach 1978) as an indication that when the channel is blocked, it cannot enter the closed state. This implies that the blocker can enter the channel only when it is open, leading to the term "open channel block" for this kind of blocking action. The total time during which the NMDA-activated channel flickered in the presence of Mg^{2+}, however, changed little or actually decreased when the degree of block became greater. This could mean that the NMDA-activated channel can close while a Mg^{2+} ion is in the channel. Another possible interpretation is that Mg^{2+} has a second inhibitory effect on the NMDA receptor; it may, for example, increase the probability of channel closure.

Comparison with Whole-Cell Data

The idea that Mg^{2+} has more than one inhibitory effect on the NMDA receptor is supported by a comparison of single channel and whole-cell data. The single channel analysis presented so far concerns the rates at which Mg^{2+} causes fast block of the NMDA-activated channel. The effects on whole-cell current cannot be simply predicted from these data because the whole-cell current does not depend solely on how frequently the channel is blocked and unblocked. Whole-cell current depends also on the length of time the channel spends in the open state and the frequency of channel opening. If the open state of the NMDA-activated channel is stabilized while Mg^{2+} is blocking, then the inhibition of whole-cell current should be less than predicted by the K_{Mg} estimated from measurements of k_+ and k_-. The time the channel spends blocked is partly compensated for by the longer time during which it flickers. The extreme of this situation would be if the channel cannot close while it is blocked, as was hypothesized for the local anesthetic block of the nicotinic ACh-activated channel. If the NMDA-activated channel closes in the same way whether or not Mg^{2+} is blocking, then the inhibition of whole-cell current should be accurately predicted from the single channel data. If Mg^{2+} has some further inhibitory action on the NMDA receptor, for example, stabilization of the closed state of the channel, then the inhibition of whole-cell currents should be greater than predicted from the single channel data.

In order to compare the blockage of whole-cell currents with that of single channel currents, Eqn. (1) must be used to predict how whole-cell, current-voltage (I−V) curves will be affected by Mg^{2+}. The apparent dissociation constant of Mg^{2+} measured in whole-cell experiments (K'_{Mg}) can be related to current in the presence of Mg^{2+} by:

$$I_{Mg} = I_o(1 + [Mg^{2+}]_o/K'_{Mg})^{-1} \tag{7}$$

where I_{Mg} is whole-cell current in the presence of Mg^{2+} and I_o is the current in the absence of Mg^{2+}. Because the whole-cell response to NMDA is reasonably linear in the absence of Mg^{2+} over the voltage range being considered here (Nowak et al. 1984; Mayer and Westbrook 1987)

$$I_o = G(V - V_r) \tag{8}$$

where G is the whole-cell conductance change induced by NMDA in the absence of Mg^{2+} and V_r the reversal potential for the NMDA response. Combining Eqs. (1) (using K'_o and V'_o to indicate that these values are from whole-cell measurements), (7) and (8),

$$I_{Mg} = G(V - V_r)(1 + [Mg^{2+}]_o/(K'_o \exp(V/V'_o)))^{-1} \tag{9}$$

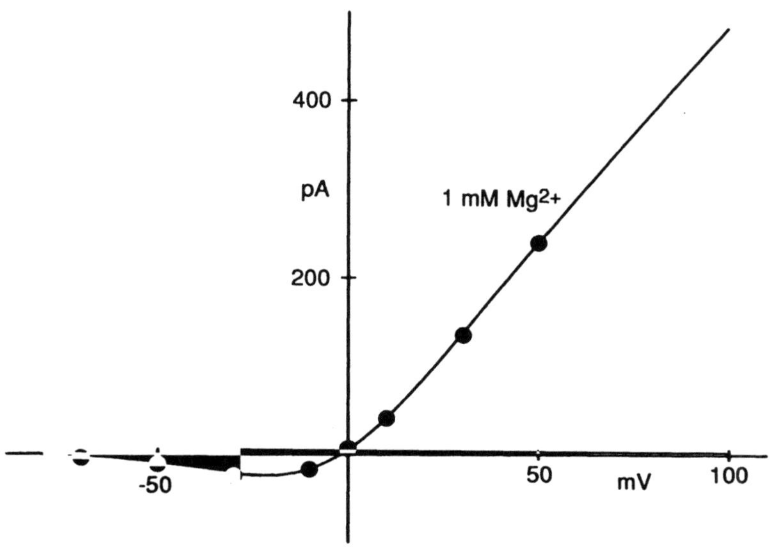

Fig. 3. I−V curve for the response to *N*-methyl-D-aspartate (NMDA) in the presence of 1 m*M* extracellular Mg^{2+}. Whole cell recording. At each voltage, 10 μ*M* NMDA + 1 μ*M* glycine was fast-perfused onto the cell (see Johnson and Ascher 1987) and the change in current measured and plotted. *Solid line* was produced by fitting Eqn. 9 to the data, leaving G, V'_o, K'_o and V_r as free parameters

This equation was fitted to a whole-cell I–V curve in the presence of 1 mM extracellular Mg^{2+} (Fig. 3). The resulting value of V_o', 15.8 mV ($\delta = 0.81$), is not very different from the value of 12.5 mV ($\delta = 1.0$) estimated from single channel measurements. It is also in reasonable agreement with the estimate made by Mayer et al. (1989) that $V_o' = 17.6$ mV ($\delta = 0.73$). Estimates from whole-cell data of K_o', however, are far different from the 8.8 mM estimate of K_o from single channel data. The value of K_o' from Fig. 3 is 1.4 mM, in good agreement with the measurement reported by Mayer et al. (1989) of 0.99 mM. These differences between whole-cell and single channel measurements of the apparent dissociation constant of Mg^{2+} from its blocking site at $V = 0$ suggest that Mg^{2+} can cause a "slow" blockage of NMDA receptor activity (Ascher and Nowak 1988) in addition to the fast blockage observed at the single channel level. The open channel block model cannot be directly applied to the Mg^{2+} block of the NMDA-activated channel. Whether the NMDA-activated channel can close while Mg^{2+} is in the channel and whether Mg^{2+} can bind to its blocking site while the channel is closed remain open questions.

Block by Intracellular Mg^{2+}

In their initial description of the blockage of the NMDA-activated channel by extracellular Mg^{2+}, Nowak et al. (1984) noted that *intracellular* Mg^{2+} also blocks the channel in a voltage-dependent manner. We have further investigated the block by intracellular Mg^{2+} (Ascher and Johnson 1988, 1989; Johnson and Ascher, 1990).

When Mg^{2+} is present in the patch pipette during single channel recording from outside-out patches, a voltage-dependent reduction in single channel current is observed (Fig. 4). The reduction is Mg^{2+} concentration dependent, and in contrast to the block by extracellular Mg^{2+}, it increases at more depolarized voltages. It is not a nonspecific effect of intracellular divalent cations, as 1 mM Ca^{2+} does not mimic the effects of 1 mM Mg^{2+} (Figs. 4, 5).

The blockage by intracellular Mg^{2+} is not associated with any significant channel flicker, unlike that by extracellular Mg^{2+} (compare Figs. 1 and 4). However, the voltage and Mg^{2+} concentration dependence of the block by intracellular Mg^{2+} suggest that a similar blocking mechanism is responsible; Mg^{2+} ions in solution enter a fraction of the membrane field to bind reversibly to a blocking site, probably within the NMDA-activated channel. We hypothesize that no flicker is seen with intracellular Mg^{2+} because the blocking and unblocking rates are very fast. Because of the limited bandwidth of our recording system (data are generally filtered at 2 kHz), a channel that is flickering with a k_+ and k_- much greater than about 10000 s^{-1} would appear to open without flicker but to pass less current. The apparent single channel current would be the average current during channel opening and would therefore reflect the fraction of time the channel spends unblocked. The apparent single channel current in the presence of $[Mg^{2+}]_i$ (i_{Mg}) is then:

$$i_{Mg} = i_o k_- (k_- + [Mg^{2+}]_i k_+)^{-1}$$

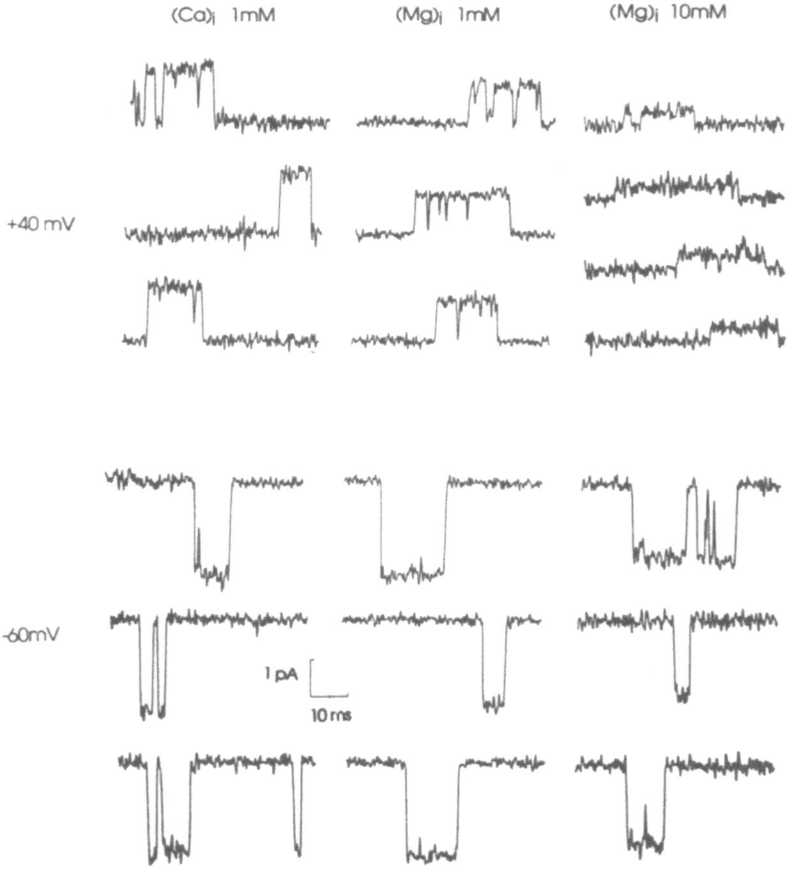

Fig. 4. Intracellular Mg^{2+} blocks the *N*-methyl-D-aspartate (NMDA)-activated channel at depolarized voltages. Data from three different outside-out patches with 1 m*M* Ca^{2+}, 1 m*M* Mg^{2+}, or 10 m*M* Mg^{2+} included in the pipette solution. *Lower half,* the current flowing through a single NMDA-activated channel can be seen to be the same with 1 m*M* Ca^{2+} or 1 m*M* Mg^{2+} in the pipette, and slightly reduced with 10 m*M* Mg^{2+}. *Upper half* the single channel current with 1 m*M* Mg^{2+} in the pipette is significantly smaller than that with 1 m*M* Ca^{2+}. When 10 m*M* Mg^{2+} is included in the pipette solution, the single channel current is further reduced. Reprinted by permission of Oxford University Press from Ascher and Johnson (1989)

where i_o is the single channel current in the absence of Mg^{2+}. Making use of Eqn. (4) and the fact that the I–V curve for single channel currents is linear in the absence of Mg^{2+} (Nowak et al. 1984; Ascher et al. 1988), Eqn. (9) can be reproduced with single channel currents and conductances replacing whole cell currents and conductances. Thus, this equation was fitted to single channel i–V curves measured in the presence of intracellular Mg^{2+} to test the idea that the same mechanism of block was at work. The data were well fit with $[Mg^{2+}]_i$

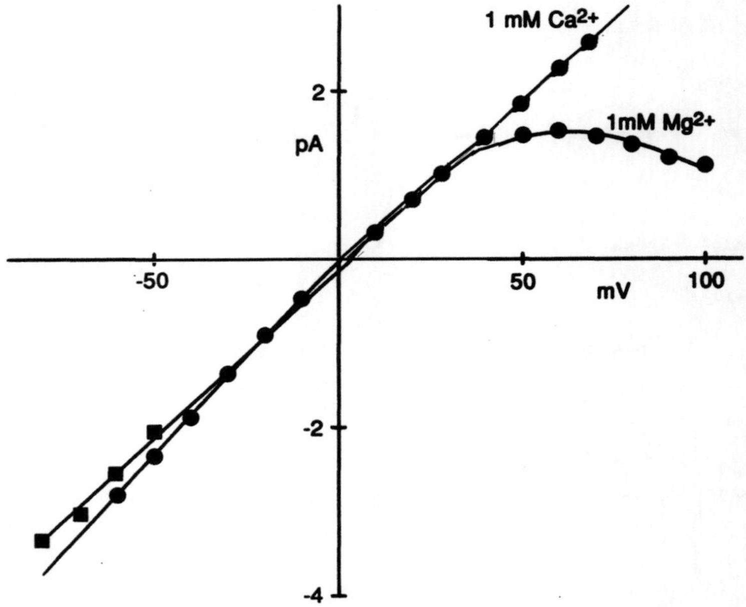

Fig. 5. The single *N*-methyl-D-aspartate (NMDA)-activated channel i–V curve. is made voltage dependent by Mg^{2+}. Data from two different outside-out patches. With 1 m*M* Ca^{2+} added to the pipette solution, the curve is nearly linear. With 1 m*M* Mg^{2+} in the pipette solution the single channel current is increasingly blocked as the patch is depolarized. *Solid line* through the 1 m*M* Ca^{2+} points is a linear least squares fit. *Solid line* through the 1 m*M* Mg^{2+} points was fitted as described in the Fig. 3 legend using Eqn. 9 for single channel data

from 0.3 to 30 m*M* and over a wide range of membrane voltages. In addition, the estimated K_o and the voltage dependence of the block were consistent at all Mg^{2+} concentrations tested. The mean value of K_o from all experiments with intracellular Mg^{2+} was 8 m*M*. The mean value of V_o was 36.8 mV, corresponding to a value of $\delta = 0.35$. Here, δ is the fraction of the membrane field felt at the blocking site measured from the intracellular side of the membrane.

Comparison with Other Types of Mg^{2+} Block

The similarity of K_o for the blocking site for intracellular and extracellular Mg^{2+} suggests that there may be only one site involved. If this were the case, then the unblocking rate from the blocking site should be the same at any given voltage, regardless of from which side the Mg^{2+} came. We found that this is not so. For example, at −40 mV, the unblocking rate of extracellular Mg^{2+} was slow enough to cause complete channel closures. However, when the channel was blocked at −40 mV by intracellular Mg^{2+}, there was a decrease in apparent channel conductance with no clear flickering. This suggests that the unblocking rate of intracellular Mg^{2+} is many times faster than that of extracellular Mg^{2+} at -40 mV. When intracellular and extracellular Mg^{2+} were applied at the same time, the two separate effects of Mg^{2+} from each side were still clearly visible. These data indicate that intracellular and extracellular Mg^{2+} block the channel by different mechanisms, probably due to binding at two different sites.

A comparison of the δs estimated for the block by intracellular and extracellular Mg^{2+} leads to a paradox. Extracellular Mg^{2+} is calculated to block at the inner membrane surface, while intracellular Mg^{2+} blocks about ⅓ of the way out from the inner membrane surface. If these figures were correct, Mg^{2+} from either side would have to pass by one binding site without action to reach its own binding site. This seems unlikely. Thus, as previously suggested, these estimates of δ probably cannot be directly applied to measure the position within the membrane field of the blocking site.

This contradiction suggests that at least one of the three assumptions required by Eqn. (2) of the model is incorrect. There are, for example, no data that clearly demonstrate the validity of assumption 3), that Mg^{2+} always unblocks to the same side of the membrane as it came from. If this is not the case, then Mg^{2+} must be able to permeate the channel, as has been suggested for extracellular Mg^{2+} at very hyperpolarized voltages by Mayer and Westbrook (1987). Because the unbinding rate constant for extracellular Mg^{2+} is increased by depolarization, extracellular Mg^{2+} must unblock most often back to the extracellular solution in the voltage range that was studied. If we suppose that Mg^{2+} sometimes unblocks in the opposite direction, then k_- and K_{Mg} should not vary single exponentially with voltage. The data, however, are consistent with a single exponential dependence. In addition, if Mg^{2+} were sometimes to unbind to the opposite side of the membrane, it would mean that we had understimated δ. The problem, however, is that they are so large that there is a crossing of the δs; this cannot be resolved by hypothesizing that Mg^{2+} can permeate the channel.

A more likely candidate for the incorrect assumption is 2), that no ions other than Mg^{2+} are involved in the block. If the access of Mg^{2+} to its blocking site involves the movement of other ions through the channel, then the effective valence (z), of Eqn. (1) would have to be increased. This could be the case if, for example, the NMDA-activated channel acts as a multi-ion single file pore (Hille and Schwarz 1978; Armstrong et al. 1982). If it were necessary to increase the valence due to the involvement of other ions, the calculated value of δ would decrease proportionally.

Several other channels are blocked by intracellular Mg^{2+} ions. The blockage of the ATP-sensitive K^+ channel (Findlay 1987; Horie et al. 1987; Ciani and Ribalet 1988), the muscarinic K^+ channel (Horie and Irisawa 1989), the voltage-activated Na^+ channel (Pusch et al. 1989), and the inwardly rectifying K^+ channel (Matsuda et al. 1987; Vandenberg 1987; Matsuda 1988) appear at least superficially similar to that of the NMDA-activated channel. The voltage dependence of the block and the apparent affinity of Mg^{2+} at 0 mV vary considerably among these channels. However, in all cases the blockage is Mg^{2+} concentration dependent, it increases with depolarization, and the data are consistent with a model in which occupation of a binding site in the channel by one or more Mg^{2+} ions results in a full and reversible block of the channel. The nicotinic ACh-receptor is also blocked by intracellular, as well as extracellular, Mg^{2+} (Dani and Eisenman 1987; Imoto et al. 1988). It is similar to the Mg^{2+} blocks of the NMDA-activated channel in that intracellular and extracellular Mg^{2+} appear to interact with the channel at two separate sites. Each site is a ring of negatively charged amino acids (Imoto et al. 1988). In addition, there is a third ring with which Mg^{2+} from either side can interact. However, this Mg^{2+}-dependent inhibition differs from that of the NMDA-activated channel in its voltage dependence. Intracellular Mg^{2+} causes a break in the i–V curve of the nicotinic channel at 0 mV such that the conductance for outward current is lower than that for inward current. The i–V curve is linear on either side of 0 mV. This suggests that the mechanism by which Mg^{2+} blocks the nicotinic ACh-receptor is distinct from that by which it blocks the NMDA receptor.

Implications

The physiological significance of the block by intracellular Mg^{2+} has not been investigated. If the normal intracellular Mg^{2+} concentration is near 1 m*M* (see Vink et al. and McGuigan et al., this volume), then the estimated apparent dissociation constant of 8 m*M* at 0 mV suggests that the block would be significant only at voltages distinctly more positive than 0 mV. During action potentials the membrane voltage would become sufficiently positive to induce a substantial block of the NMDA response. In the absence of any blockage by intracellular Mg^{2+}, an NMDA response could considerably shunt the action potential. It is possible that the blockage by intracellular Mg^{2+} serves to reduce this shunt.

Our estimate of the apparent dissociation constant of intracellular Mg^{2+} from the NMDA receptor depended solely on single channel measurements. It may be that intracellular Mg^{2+} has a second blocking action that was not apparent from single channel measurements, as appears to be the case for extracellular Mg^{2+}. If this is so, the K_o may be in effect lower. The results described by Vink et al. (this volume) suggest that an association may exist between a drop in intracellular Mg^{2+} concentration following brain trauma and tissue damage that may in part be due to NMDA receptor overactivation. If the effective K_o' is much below 8 m*M*, it is plausible that under conditions in which neurons are depolarized, intracellular Mg^{2+} could significantly block NMDA responses.

Conclusions

The NMDA receptor channel complex is highly regulated by Mg^{2+} ions. Extracellular Mg^{2+} blocks the channel in a voltage-dependent manner, probably by entering the channel and preventing ion flow. The blockage is increased by hyperpolarization. Extracellular Mg^{2+} may have more than one inhibitory effect. The resulting nonlinearity in the I–V curve of the NMDA response is of great functional importance in synaptic transmission and its regulation.

Intracellular Mg^{2+} also causes a voltage-dependent blockage of the NMDA response. At 0 mV the apparent dissociation constant, measured in single channel experiments, is about the same for intracellular and extracellular Mg^{2+}. The block by intracellular Mg^{2+} is not as steeply voltage dependent, it is increased by depolarization, and the two kinds of block are significant on opposite sides of 0 mV. The kinetics of the block by intracellular Mg^{2+} is much faster than that by extracellular Mg^{2+}. Different blocking sites are used by Mg^{2+} from different sides of the membrane.

The NMDA receptor is thus a bidirectional rectifier. Under physiological conditions, it passes current well only in a relatively restricted voltage region centered near 0 mV.

References

Armstrong CM, Swenson RP, Taylor SR (1982) Block of squid axon K channel by internally and externally applied barium ions. J Gen Physiol 80:663–682

Ascher P, Johnson JW (1988) Effects of external and internal Mg on the NMDA-activated channel. Biol Bull 175:433

Ascher P, Johnson JW (1989) The NMDA receptor, its channel, and its modulation by glycine. In: Watkins J, Collingridge G (eds) The NMDA receptor. Oxford University Press, Oxford, pp 109–121

Ascher P, Nowak L (1988) The role of divalent cations in the N-methyl-D-aspartate responses of mouse central neurones in culture. J Physiol (Lond) 399:247–266

Ascher P, Bregestovski P, Nowak L (1988) N-methyl-D-aspartate-activated channels of mouse central neurones in magnesium free solutions. J Physiol (Lond) 399:207–226

Ault B, Evans RH, Francis AA, Oakes DJ, Watkins JC (1980) Selective depression of excitatory amino acid induced depolarizations by magnesium ions in isolated spinal cord preparations. J Physiol (Lond) 307:413–428

Bliss TVP, Lynch MA (1988) Long-term potentiation of synaptic transmission in the hippocampus: properties and mechanisms. In: Landfield PW, Deadwyler SA (eds) Long-term potentiation: from biophysics to behavior. Liss, New York, pp 3–72

Choi DW (1988) Glutamate neurotoxicity and diseases of the nervous system. Neuron 1:623–634

Ciani S, Ribalet B (1988) Ion permeation and rectification in ATP-sensitive channels from insulin-secreting cells (RIN m5F): effects of K^+, Na^+ and Mg^{2+}. J Membr Biol 103:171–180

Cline HT, Debski EA, Constantine-Paton M (1987) N-Methyl-D-aspartate receptor antagonist desegregates eye-specific stripes. Proc Natl Acad Sci USA 84:4342–4345

Dani JA, Eisenmann G (1987) Monovalent and divalent cation permeation in acetylcholine receptor channels. Ion transport related to structure. J Gen Physiol 89:959–983

Findlay I (1987) The effects of magnesium upon adenosine triphosphate-sensitive potassium channels in a rat insulin-secreting cell line. J Physiol (Lond) 391:611–629

Hamill OP, Marty A, Neher E, Sakmann B, Sigworth FJ (1981) Improved patch-clamp techniques for high resolution current recording from cells and cell-free membrane patches. Pflugers Arch 391:85–100

Herron CE, Lester RAJ, Coan EJ, Collingridge GL (1986) Frequency-dependent involvement of NMDA receptors in the hippocampus: a novel synaptic mechanism. Nature 322:265–268
Hille B, Schwarz W (1978) Potassium ions as multi-ion single file pores. J Gen Physiol 72:409–442
Horie M, Irisawa H (1989) Dual effects of intracellular magnesium on muscarinic potassium channel current in single guinea-pig atrial cells. J Physiol (Lond) 408:313–332
Horie M, Irisawa H, Noma A (1987) Voltage-dependent magnesium block of adenosine-triphosphate-sensitive potassium channel in guinea-pig ventricular cells. J Physiol (Lond) 387:251–272
Imoto K, Busch C, Sakmann B, Mishina M, Konno T, Nakai J, Bujo H, Mori Y, Fukuda K, Numa S (1988) Rings of negatively charged amino acids determine the acetylcholine receptor conductance. Nature 335:645–648
Johnson JW, Ascher P (1987) Glycine potentiates the NMDA response in cultured mouse brain neurons. Nature 325:529–531
Johnson JW, Ascher P (1990) Voltage-dependent block by intracellular Mg^{2+} of *N*-methyl-D-aspartate-activated channels. Biophys J 57:1085–1090
McDermott AB, Mayer ML, Westbrook GL, Smith SJ, Barker JL (1986) NMDA-receptor activation increases cytoplasmic calcium concentration in cultured spinal cord neurones. Nature 321:519–522
Matsuda H (1988) Open-state substructure of inwardly rectifying potassium channels revealed by magnesium block in guinea-pig heart cells. J Physiol (Lond) 397:237–258
Matsuda H, Saigusa A, Irisawa H (1987) Ohmic conductance through the inwardly rectifying K^+ channel and block by internal Mg^{2+}. Nature 325:156–159
Mayer ML, Westbrook GL (1987) Permeation and block of *N*-methyl-D-aspartic acid receptor channels by divalent cations in mouse cultured central neurons. J Physiol (Lond) 394:501–527
Mayer ML, Westbrook GL, Guthrie PB (1984) Voltage-dependent block by Mg^{2+} of NMDA responses in spinal cord neurones. Nature 309:261–263
Mayer ML, Vyklicky L Jr, Westbrook GL (1989) Modulation of excitatory amino acid receptors by group II B metal cations in cultured mouse hippocampal neurones. J Physiol (Lond) 415:329–350
Neher E, Steinbach JH (1978) Local anaesthetics transiently block currents through single acetylcholine-receptor channels. J Physiol (Lond) 277:153–176
Nowak L, Bregestovski P, Ascher P, Herbet A, Prochiantz A (1984) Magnesium gates glutamate-activated channels in mouse central neurones. Nature 307:462–465
Pusch M, Conti F, Stuhmer W (1989) Intracellular magnesium blocks sodium outward currents in a voltage- and dose-dependent manner. Biophys J 55:1267–1271
Vandenberg CA (1987) Inward rectification of a potassium channel in cardiac ventricular cells depends on internal magnesium ions. Proc Natl Acad Sci USA 84:2560–2564
Woodhull AM (1973) Ionic blockage of sodium channels in nerve. J Gen Physiol 61:687–708

Excess Mg^{2+} and Central Nervous System Metabolism

Gregory Crosby [1] and Michele D. Szabo [1]

There is perhaps no more dramatic evidence for a depressant effect on the central nervous system (CNS) than the ability of a drug or compound to produce a state of general anesthesia. Mg^{2+} was at one time thought to have such marked effects on the CNS that early in this century Peck and Meltzer (1916) reported performing surgery on humans under Mg^{2+} "anesthesia." Less profound CNS signs and symptoms, ranging from drowsiness to light coma, have also been reported in hypermagnesemic individuals (Randall et al. 1964; Mordes and Wacker 1978; Paymaster 1976).

However, in careful experiments on human volunteers, Somjen et al. (1966) concluded that a plasma Mg^{2+} level of approximately 15 mEq/l does not produce sedation, let alone general anesthesia. In fact, they reported that the volunteers (two of the study's authors), despite being paralyzed because of the effect of Mg^{2+} on neuromuscular transmission (Mordes and Wacker 1978), remained conscious and cognizant of pain. How can one expect Mg^{2+} to be a CNS depressant if it does not produce general anesthesia or even sedation?

Part of the answer lies in biochemical and neurophysiological data. For instance, a large body of data indicates that Mg^{2+} is a critical cofactor for many enzyme systems. One of its main effects is the activation of enzymes (Paymaster 1976; Ebel and Gunter 1980). Approximately 300 enzymes require Mg^{2+} for activation (Ebel and Gunter 1980), a major group being those involved in reactions with adenosine triphosphate. Since adenosine triphosphate is necessary for numerous biochemical reactions, including glucose utilization and protein synthesis, the activating effect of Mg^{2+} becomes important for all of these functions (Paymaster 1976; Ebel and Gunter 1980). Mg^{2+} can also inhibit enzyme activity. Thus, the catalytic activity of Mg^{2+}-dependent enzymes is reduced at high Mg^{2+} concentrations (Ebel and Gunter 1980), evidently as a result of non-specific binding of Mg^{2+} to the enzyme protein. One example of this is its ability to inhibit phosphofructokinase (Bachelard and McIlwain 1985), one of the rate-limiting enzymes of glycolysis. Since the CNS depends almost exclusively on glucose for energy metabolism (Sokoloff 1981), it would not be surprising if excess Mg^{2+} reduced glucose utilization in the brain.

---

[1] Anesthesia Services of the Massachusetts General Hospital, Boston MA 02114, and the Department of Anaesthesia, Harvard Medical School, Boston, MA 02115, USA

Decreases in CNS metabolism can also occur by mechanisms other than direct effects on important enzyme systems. The metabolic rate of CNS tissue is in large measure determined by and very closely reflects the functional activity of the tissue (Sokoloff 1977). Therefore, by changing neuronal activity, one can change the metabolic rate. Mg^{2+} has a variety of effects that may reduce neural activity. Excess Mg^{2+} affects the peripheral nervous system by reducing the release of acetylcholine at the neuromuscular junction and by decreasing the responsiveness of the postsynaptic muscle membrane (Del Castillo and Engbaek 1954). The resulting block of transmission at the neuromuscular junction produces paralysis of voluntary muscles at plasma Mg^{2+} concentrations above 10 mEq/l (Mordes and Wacker 1978; Ebel and Gunter 1980). A reduction in proprioceptive input to the CNS resulting from muscle paralysis could have secondary effects on the brain's metabolic rate. CNS depression could also theoretically occur from peripheral, nonspecific effects of excess Mg^{2+}, such as hypoxia (due to untreated neuromuscular paralysis) or hypotension, and some have speculated that these systemic effects are responsible for the "anesthesia" sometimes attributed to excess Mg^{2+}.

Within the CNS, however, direct application of Mg^{2+} blocks synaptic transmission (Kato and Somjen 1969; Somjen and Kato 1968; Kato et al. 1968; Wojtowicz et al. 1977; Borges and Gucer 1978). For example, microiontophoretic application of Mg^{2+} ions onto neurons in the cerebral cortex, cuneate nucleus, and spinal cord profoundly suppresses spontaneous and glutamate- or acetylcholine-stimulated firing (Somjen and Kato 1968). This effect does not appear to be entirely presynaptic, however, since Mg^{2+} increases the threshold to both electrical (Somjen and Kato 1968; Wojtowicz et al. 1977) and neurochemical (Kato et al. 1968) stimulation of neurons with little or no change in the spontaneous miniature endplate potentials.

There are, surprisingly, few studies in which the CNS metabolic effects of Mg^{2+} have been measured directly. Ames and Nesbett (1983) recently used an isolated rabbit retina preparation and biochemical analysis of 2-[^{14}C]-deoxyglucose (2-[^{14}C]DG) uptake to evaluate the metabolic effects of excess Mg^{2+}. In that in vitro model, increasing the Mg^{2+} concentration in the incubation medium from 1.2 to 15 m*M* resulted in cessation of spontaneous and light-evoked activity in the retinal neurons and a 68% decrease in 2-[^{14}C]DG uptake. Those results are consistent with the hypothesis that Mg^{2+} is a CNS depressant that acts in part by decreasing neuronal activity. Reports of its in vivo metabolic effects are not so consistent, however.

The earliest report (McCall and Sass 1956) involved 15 patients with toxemia of pregnancy who received 20 ml of a 10% solution (approximately 2 g) of magnesium sulfate intravenously. Cerebral blood flow (CBF) and cerebral metabolic rate for oxygen ($CMRO_2$) were measured with the nitrous oxide technique. Rather than decreasing metabolism, Mg^{2+} was found to increase both $CMRO_2$ (3.8 vs. 3.6 ml/100 g per min in control state, $P<0.01$) and CBF. The fact that regional metabolism was not assessed and plasma Mg^{2+} concentrations not reported makes the study difficult to interpret. Indeed, inasmuch as only a single, relatively small dose of Mg^{2+} was given, it seems likely that the patients were studied during only mild to moderate hypermagnesemia. In more recent work, the effect of maternal hypermagnesemia on fetal brain blood flow and metabolism was measured in sheep using electromagnetic flow probes and arterial-venous oxygen content differences (Ayromlooi et al. 1982). Although the maternal serum Mg^{2+} concentration increased from 2.5 to approximately 8 mg%, the fetal Mg^{2+} level did not change. Fetal brain oxygen consumption decreased transiently to about 75% of control but returned to normal within 60 min. Since the level of fetal plasma Mg^{2+} did not increase, it is interesting that fetal brain oxygen consumption changed at all.

On the other hand, our work (Szabo and Crosby 1988) indicates that profound hypermagnesemia does produce a substantial decrease in CNS metabolism. In that study, spinal glucose utilization was measured autoradiographically with the 2-[^{14}C]DG method in mechanically ventilated rats breathing 70% nitrous oxide. Magnesium sulfate was delivered by continuous intravenous infusion beginning 30 min prior to the experiment, and plasma Mg^{2+} concentrations reached 6.75 ± 0.5 m*M* (vs. 0.9 ± 0.5 m*M* in controls; $P < 0.01$). Under these conditions, spinal glucose utilization decreased 26%–45% in gray and 53%–63% in white matter. This metabolic response is similar to that reported by Ames and Nesbett (1983) in in vitro experiments in retina. The disproportionately larger effect of Mg^{2+} on the white matter is unusual because white matter has a much lower rate of metabolism than gray matter and typically is affected less by metabolic depressants (Crosby et al. 1984) or events that block impulse conduction (Crosby 1985; Schwartzman et al. 1983). Moreover, the mechanism for these effects is uncertain. Although the Mg^{2+}-treated animals were hypotensive (88 ± 1 vs. 130 ± 4 mmHg; $P < 0.01$) (Szabo and Crosby 1988), this is unlikely to have affected the results because blood pressure remains within autoregulatory limits (Sandler and Tator 1976), and spinal metabolic rate is unaffected by similar blood pressure changes in anesthetized animals (Crosby and Atlas 1988). Thus, systemic physiological effects of Mg^{2+} do not provide a satisfactory explanation for metabolic depression. Although peripheral effects such as reduced conduction along the sensory limb of peripheral nerves and/or loss of proprioceptive signals are plausible explanations, central effects of Mg^{2+} on glycolysis and synaptic transmission, such as those mentioned previously, would seem the most reasonable cause of such marked metabolic changes. There is controversy as to whether Mg^{2+} levels in the CNS increase sufficiently during hypermagnesemia to account for the substantial metabolic changes reported, however.

Although under normal circumstances the Mg^{2+} concentration in the cerebrospinal fluid (CSF) is significantly greater than in plasma (Oppelt et al. 1963), it is equally clear that uptake of Mg^{2+} from the blood into the CSF and brain is limited (Oppelt et al. 1963; Hilmy and Somjen 1968; Cserr 1971). The limitation does not appear to be a simple matter of relative blood-brain barrier impermeability; evidence indicates that an active transport system carefully regulates the CSF/plasma Mg^{2+} ratio (Oppelt et al. 1963). Changes in the CSF Mg^{2+} level do nevertheless occur during hypermagnesemia, but the increases are much less than one might anticipate from the plasma levels. For example, in a toxemic patient whose plasma Mg^{2+} concentration was maintained between 6.3 and 11 mEq/l for 7 days, the CSF Mg^{2+} concentration increased about 1 mEq/l (Pritchard 1955). In experiments performed by Oppelt (1963), administration of sufficient Mg^{2+} to raise the plasma concentration of 5 mEq/l produced a 21% increase in the CSF concentration. Hilmy and Somjen (1968) also reported that intravenously administered Mg^{2+} increased brain tissue uptake by about 0.8 mEq/kg fresh tissue, an increase far less than that seen in peripheral nerve and muscle. Interestingly, Mg^{2+} uptake was approximately 30% higher in the spinal cord than in the cortex, diencephalon, or midbrain. In this regard, it is noteworthy that during Mg^{2+} depletion loss of Mg^{2+} from the white matter exceeds that from the gray matter (Chutkow 1974). Thus, the Mg^{2+} concentration of the white matter appears to be more sensitive to changes in plasma Mg^{2+}, and this observation may account for the greater metabolic effect of hypermagnesemia on spinal white matter tracts.

The key issue is whether such relatively small increases in Mg^{2+} concentration are adequate to produce biological effects. The answer is unclear, but some data suggest that small changes in Mg^{2+} are associated with significant alterations in brain function. For instance, from microiontophoretic studies, Somjen and Kato (1968) calculated that the extracellular fluid Mg^{2+} concentration would have to increase only 1 or at most 2 mEq/l to produce marked suppression of neuronal firing. Even following severe traumatic brain injury in rats, the free Mg^{2+} level decreases only about 0.6 m*M* (Vink et al. 1987), and it has been hypothesized that such a decrease may be a critical factor in the development of irreversible tissue injury.

Thus, while excess Mg^{2+} does not produce an anesthetic state, it does reduce neuronal activity and metabolic rate in the CNS. There is little question that its effects are profound when applied directly onto nervous tissue, such as during in vitro experiments or perhaps when blood-brain barrier or transport system function are impaired. Contrary to some studies, recent work indicates that hypermagnesemia reduces glucose utilization even in intact, normal animals. This effect is most prominent in the white matter, perhaps because Mg^{2+} concentrations in that tissue increase somewhat more than in the gray matter. While the mechanism for the metabolic effects of parenteral Mg^{2+} is somewhat speculative, the most obvious explanation is that the relatively small increases in the CSF/brain Mg^{2+} concentration that do occur are nevertheless sufficient to produce substantial changes in brain energy metabolism.

References

Ames A III, Nesbett FB (1983) Pathophysiology of ischemic cell death. I. Time of onset of irreversible damage: importance of the different components of the ischemic insult. Stroke 14:219–226

Ayromlooi J, Desiderio DM, Tobias M, Berg P (1982) Effect of magnesium sulfate on maternal and fetal hemodynamics and fetal brain function and metabolism. Pediatr Pharmacol 2:305–315

Bachelard HS, McIlwain H (1985) Biochemistry and the central nervous system. Churchill Livingstone, New York, pp 131–153

Borges LF, Gucer G (1978) Effect of magnesium on epileptic foci. Epilepsia 19:81–90
Chutkow JG (1974) Metabolism of magnesium in the central nervous system. Neurology 24:780–787
Crosby G (1985) Local spinal cord blood flow and glucose utilization during spinal anesthesia with bupivacaine in conscious rats. Anesthesiology 63:55–60
Crosby G, Atlas S (1988) Local spinal cord glucose utilization in conscious and halothane-anesthetized rats. Can J Anaesth 35:359–363
Crosby G, Crane AM, Sokoloff L (1984) A comparison of local rates of glucose utilization in spinal cord and brain in conscious and nitrous oxide- or pentobarbital-treated rats. Anesthesiology 61:434–438
Cserr HF (1971) Physiology of the choroid plexus. Physiol Rev 51:273–311
Del Castillo J, Engbaek L (1954) The nature of the neuromuscular block produced by magnesium. J Physiol (Lond) 124:370–384
Ebel H, Gunter T (1980) Magnesium metabolism: a review. J Clin Chem Clin Biochem 18:257–270
Hilmy MI, Somjen GG (1968) Distribution and tissue uptake of magnesium related to its pharmacological effects. Am J Physiol 214:406–413
Kato G, Somjen GG (1969) Effect of micro-iontophoretic administration of magnesium and calcium on neurons in the central nervous system of cats. J Neurobiol 2:181–195
Kato G, Kelly JS, Krnjevic K, Somjen G (1968) Anesthetic action of magnesium ions. Can Anaesth Soc J 15:539–544
McCall ML, Sass D (1956) The action of magnesium on cerebral circulation and metabolism in toxemia of pregnancy. Am J Obstet Gynecol 71:1089–1096
Mordes JP, Wacker WC (1978) Excess magnesium. Pharmacol Rev 29:273–300
Oppelt WW, MacIntyre I, Rall DP (1963) Magnesium exchange between blood and cerebrospinal fluid. Am J Physiol 205:959–962
Paymaster NJ (1976) Magnesium metabolism: a brief review. Ann R Coll Surg Engl 58:309–314
Peck CH, Meltzer SJ (1916) Anesthesia in human beings by intravenous injection of magnesium sulfate. JAMA 67:1131–1133
Pritchard JA (1955) The use of magnesium ion in the management of eclamptogenic toxemias. Surg Gynecol Obstet 100:131–140
Randall RE, Cohen D, Spray CC, Rossmeisel C (1964) Hypermagnesemia in renal failure. Ann Intern Med 61:73–88
Sandler AN, Tator CH (1976) Review of the measurement of normal spinal cord blood flow. Brain Res 118:181–196
Schwartzman RJ, Eidelberg E, Alexander GM, Yu J (1983) Regional metabolic changes in spinal cord related to spinal shock and later hyperreflexia in monkeys. Ann Neurol 14:33–37
Sokoloff L (1977) Relation between physiological function and energy metabolism in the central nervous system. J Neurochem 29:13–26
Sokoloff L (1981) Circulation and energy metabolism of the brain. In: Siegel GJ, Albers RW, Agranoff BW, Katzman R (eds) Basic neurochemistry, 3rd edn. Little, Brown, Boston, pp 471–495
Somjen G, Kato G (1968) Effects of magnesium and calcium on neurons in the central nervous system. Brain Res 9:161–164
Somjen G, Hilmy M, Stephen CR (1966) Failure to anesthetize human subjects by intravenous administration of magnesium sulfate. J Pharmacol Exp Ther 154:652–659
Szabo MD, Crosby G (1988) Effect of profound hypermagnesemia on spinal cord glucose utilization in rats. Stroke 19:747–749
Vink R, McIntosh TK, Demeduik P, Faden AI (1987) Decrease in total and free magnesium concentration following traumatic brain injury in rats. Biochem Biophys Res Commun 149:594–599
Wojtowicz JM, Marshall KC, Hendelman WJ (1977) Depression by magnesium ion of neuronal excitability in tissue cultures of central nervous system. Can J Physiol Pharmacol 55:367–372

Mg^{2+} in Neurotrauma: Its Role and Therapeutic Implications

Robert Vink[1], Tracy K. McIntosh[2], and Alan I. Faden[3]

Introduction

In Western, developed countries such as the USA, UK, New Zealand, and Australia, injury to the brain or spinal cord results in more deaths in individuals under 44 years of age than all other causes of death (Selecki et al. 1980; Frankowski et al. 1985). As such it can be considered the primary killer of young people in these countries. Aside from the significant mortality and morbidity associated with central nervous system (CNS) injury, in financial terms the cost to the community for hospitalization and rehabilitation runs into billions of dollars per year. Recent developments suggest that much of the injury associated with CNS trauma can be prevented. While some of the tissue damage associated with the development of irreversible injury occurs at the time of injury, much of it is delayed, arising hours to days after the initial insult (Cooper 1985). Recognition of the factors that contribute to the development of irreversible injury permits development of "antifactors" that may attenuate or even prevent manifestation. This review will concentrate on one factor that has been receiving increasing attention in the study of neurotrauma: Mg^{2+}.

Neurotrauma

Neurotrauma can be described as traumatic injury to the brain or spinal cord, often resulting in irreversible tissue injury with resultant neurological deficit. Two mechanisms have been identified as contributing to the development of irreversible tissue damage following neurotrauma. At the time of the trauma, primary events may contribute to injury by mechanical phenomena such as neuronal shearing, transection, and axonal stretching (Povlishock 1985; Noble and Wrathall 1989). After the primary event, secondary events are initiated that manifest over a period of hours to days following the insult (Cooper 1985). These include physiological and biochemical events such as reduced

[1] Department of Chemistry and Biochemistry, James Cook University of North Queensland, Townsville, Queensland 4811, Australia
[2] Department of Surgery, University of Connecticut Health Center, Farmington, CT, USA
[3] Department of Neurology, University of California School of Medicine, and Veterans Administration Medical Center, San Francisco, CA, USA

blood flow (Sandler and Tator 1976), edema (Tornheim 1985; Stewart 1985), ion changes (Young 1985 b; Vink et al. 1988 b; McIntosh and Vink 1989), loss of energy reserves (Walker et al. 1977; Vink et al. 1989 a), membrane breakdown (Ellis et al. 1981; Demediuk et al. 1985; Siesjo and Wieloch 1985), and release of "autodestructive" agents including opioid peptides (Faden et al. 1981) and excitatory amino acids (Faden and Simon 1988; Hayes et al. 1988; Faden et al. 1989). While little can be done therapeutically with respect to primary events, the fact that secondary injury develops at delayed time points following the initial trauma has been the impetus to identify the factors involved and to develop therapies aimed at preventing or attenuating their destructive effects.

Several models have been developed to study trauma to the brain and spinal cord (Gennarelli and Thibault 1985; Faden 1985). Currently, the one most widely used for brain injury is the fluid percussion model developed by Sullivan et al. (1976) in the cat, which was later further developed and adapted to other species (Dixon et al. 1987; McIntosh et al. 1989 b). This model takes advantage of a carefully calibrated saline pressure pulse injected epidurally into the closed cranial cavity causing transient brain deformation and resultant injury. The pressure pulse can be varied in intensity; however, the duration of the pressure pulse remains constant (approximately 22 ms). The neurological outcome is assessed using a battery of motor tests and the animals classified according to performance. The advantages of this model are that is is highly reproducible, simulates many of the physiological and histological consequences of clinical brain injury, and can easily be regulated to give varying neurological outcomes. The disadvantages are that it does not reproduce all of the physiological and morphological changes noted in clinical brain injury, and that the titration curve to severe injury is so narrow that experimental injury is realistically limited to studies of mild and moderate injury. Variations of this model have been developed which address some of these issues (McIntosh et al. 1989 b).

The most widely used models of traumatic spinal cord injury are the Allen weight drop method (Allen 1911) and the compression model (Tarlov 1972; Rivlin and Tator 1978). The former involves dropping a weight from a predetermined height through a guide tube onto an impounder resting on the exposed spinal cord. The impact is recorded as a gm·cm injury (multiplying the weight used by the height of drop) and is classified as mild, moderate, or severe based on the motor outcome. The compression model involves sitting a weight on the exposed spinal cord for a defined length of time, or attaching a clip to the spinal cord for a defined length of time. All of these models have been successfully used to characterize a variety of secondary events; studies of Mg^{2+} concentration have been limited to the Allen weight drop method in rats. Accordingly, with respect to spinal cord injury, this review will be limited to this model.

Mg^{2+} in the Central Nervous System

Mg^{2+} is one of the most abundant ions in mammalian tissue, being fourth in concentration behind Ca^{2+}, K^+, and Na^+, and the second most abundant in the intracellular space. It is distributed between three main compartments. Some 65% is found in the mineral phase of the hard tissues such as bone. This fraction can be readily mobilized to compensate for a plasma Mg^{2+} deficiency. The remainder is distributed between the intracellular and extracellular compartments. In the CNS, the total Mg^{2+} concentration varies from 17–22 µmoles/g dry weight in white matter to 29–33 µmoles/g dry weight in cortical tissue. Most of this is bound to metabolites, proteins, and membrane phospholipids including the polyphosphoinositides. Brain intracellular free Mg^{2+} represents 8%–10% of the total cellular Mg^{2+} pool and is of the order of 0.8–1.0 m*M* (Veloso et al. 1973; Ebel and Gunther 1980; Aikawa 1981; Vink et al. 1987a). The concentration in the cerebrospinal fluid (CSF) is 1.2 m*M*. The similarity in concentration between the intracellular and extracellular compartments suggests that there is no gradient across the cell membrane.

Over 300 enzymes require Mg^{2+} for optimal function (Heaton 1980; Ebel and Gunther 1980). Of these, one of the most important classes in terms of energy metabolism is that of the transphosphorylase enzymes involved in the transfer of phosphate groups. These enzymes are found in virtually all cell metabolic pathways including the glycolytic pathway, tricarboxylic acid cycle, and oxidative phosphorylation. All ATP-consuming and -producing reactions require Mg^{2+} for activation, with the Mg^{2+}-nucleotide complex serving as the activated substrate. Included in this group of enzymes is the Na^+/K^+-ATPase, phosphofructokinase, and creatine kinase. Other enzymes must bind directly with Mg^{2+} for activation, including pyruvate kinase, enolase, and adenylate kinase.

Mg^{2+} is required for membrane stability also (Bara and Guiet-Bara 1984). Being an integral part of the cell membrane, it regulates membrane fluidity and, perhaps indirectly, influences the activity of membrane-associated enzymes (Ebel and Gunther 1980). Its other membrane-associated functions include regulation of permeability toward other ions, notably Ca^{2+} and K^+ (Bara and Guiet-Bara 1984; B. M. Altura and B. T. Altura 1985b). Mg^{2+} has also been implicated in regulation of the cell cycle through its effects on DNA synthesis and RNA aggregation (Rubin 1976). Optimal levels of Mg^{2+} are required for both cell functions. Protein synthesis is also dependent on optimal Mg^{2+} concentrations (Gunther et al. 1984).

Physiologically, Mg^{2+} has three critical functions that are of importance to CNS function. Firstly, it regulates both neuromuscular and vascular tone. Increasing its plasma concentration dilates blood vessels, presumably through actions on Ca^{2+} channels (B. T. Altura and B. M. Altura 1982). Conversely, reduction of its plasma concentration has been shown to increase vascular tone and even to induce arterial vasospasm (B. M. Altura et al. 1984). Mg^{2+} also regulates the binding of certain receptor groups to their binding sites, notably the opiate receptors (Sadee et al. 1982), and reportedly gates the ion channel of the *N*-methyl-D-aspartate (NMDA) class of excitatory amino receptors (Nowak et al. 1984). The importance of these and other factors in secondary injury have been extensively reviewed elsewhere (Cooper 1985; Faden 1985; Siesjo and Wieloch 1985). However, a brief description of some of these factors is relevant to the discussion of the role of Mg^{2+} in neurotrauma.

Secondary Injury

Lactate and pH

The observation in clinical head injury that lactate concentrations are elevated in the CSF following trauma suggested that lactate production in injured tissue may be one factor that results in irreversible damage (Enevoldsen et al. 1976; Gulati et al. 1980; Rosner and Becker 1984; DeSalles et al. 1986). Several studies have examined the role of lactate in experimental neurotrauma (Nilsson and Ponten 1977; Wagner et al. 1985; Yang et al. 1985). The most recent brain injury experiments utilized magnetic resonance spectroscopy (MRS) to monitor not only the concentration but also the kinetics of accumulation (McIntosh et al. 1987). They used the lateral fluid percussion brain injury model to demonstrate that lactate accumulation following brain injury is transient, with concentrations never exceeding relatively low levels (i.e., 6–8 m*M*). The accumulation of lactate was linearly correlated with brain acidosis, also a transient phenomenon, pH normalizing by 90 min postinjury. Notably, these studies demonstrated that lactic acidosis was not correlated with neurological outcome.

The analyses of pH and lactate accumulation following spinal cord injury have shown striking differences from those of head injury (Walker et al. 1977). The most recent research used MRS technology to examine kinetics in vivo (Vink et al. 1989 a) and demonstrated that even at the lowest level of injury, profound lactic acidosis was associated with spinal cord trauma. The degree of acidosis was again independent of injury severity, although the rate of acidosis seemed to differ according to injury level. Concentrations of lactate were estimated to reach a maximum of 16–20 m*M* between 2 and 3 h postinjury.

Bioenergetics

Decreased bioenergetic capacity has been implicated as a contributory factor in neurotrauma (Yang et al. 1985; Vink et al. 1987 b). Reports have appeared demonstrating that alterations in high energy phosphates occur following both brain and spinal cord injury (Walker et al. 1977; Gulati et al. 1980; Yang et al. 1985; Vink et al. 1987 b, 1988 d, 1989 a). However, unlike spinal cord injury which demonstrates a profound loss of ATP following trauma (Vink et al. 1989 a), brain injury, at least in the fluid percussion model, does not exhibit significant changes in ATP concentration (Vink et al. 1987b, 1988d). Nonetheless, Vink et al. (1988a) have demonstrated that the bioenergetic capacity is compromised following brain injury. All levels of brain trauma resulted in a decreased cytosolic phosphorylation potential. While implying that this is associated with the development of irreversible tissue injury, no correlation between decreased phosphorylation potential and neurological outcome could be found. Recent studies examining mitochondrial function following brain trauma have shown that mitochondrial function is essentially preserved following trauma (R. Vink et al., unpublished results), an observation similar to that made following transient ischemia (Hillered et al. 1984). Unlike ischemia studies, however, there was no change in the coupling between respiration and ATP synthesis.

Blood Flow and Edema

Studies of blood flow following traumatic brain injury have shown that any changes in flow are focal and slight (Yuan et al. 1988; Yamakami and McIntosh 1989). Maximal changes of up to 50% reduction have been reported, but even this did not result in a reduction of blood flow to below ischemic thresholds, i.e., below 20 ml/100 g·min. This observation is supported by both MRS results examining phosphate metabolism and electrophysiological studies (Vink et al. 1987b). In contrast, blood flow reductions to below ischemic thresholds have been noted in traumatic spinal cord injury (Sandler and Tator 1976; Young 1985a). In the weight drop model, marked reductions in blood flow develop over the first 2 h posttrauma. After 2 h, the reduction in blood flow is such that both phosphate metabolism (Vink et al. 1989a) and electrophysiological function (Young et al. 1981) are severely compromised.

Increased concentration of water in CNS tissue is a common observation under conditions of reduced energy metabolism (Siesjo 1978). It is not surprising, therefore, that traumatic spinal cord injury is associated with the development of edema as early as 3 h posttrauma (Stewart 1985; Lemke et al. 1987), although somewhat unexpectedly, development of edema was independent of injury severity (Kwo et al. 1989; Lemke and Faden 1990). Traumatic brain injury on the other hand is not accompanied by edema, at least not in the acute phase (Tornheim 1985; Vink et al. 1987b). There have been reports of edema development at later time points following brain injury which can be successfully treated by therapeutic intervention (McIntosh et al. 1989a). The benefits of such treatment to neurological outcome were questionable (McIntosh et al. 1989c). One can therefore speculate that the appearance of edema in clinical brain trauma may be associated with causes other than the brain trauma per se.

Opioid Peptides and Amino Acids

A role for opioid peptides in neurotrauma was proposed by Faden and colleagues (Faden et al. 1981, 1985) on the basis of increased opioid immunoreactivity at the trauma site following spinal cord injury and the protective effects of opiate antagonists on the neurological outcome. More recent studies have demonstrated that the use of opiate antagonists is also beneficial following traumatic brain injury (Hayes et al. 1983; McIntosh et al. 1988 b) and that the deleterious effects of opioid peptides may be specifically related to the actions of a subgroup of opiate receptors, namely the $\varkappa$ opiate receptors (Faden et al. 1988 c, 1990). Perhaps of greatest interest is the recent suggestion that the secondary injury effects of opioid peptides involve nonopioid mechanisms including the excitatory amino acid receptors (Caudle and Isaac 1988; Bakshi and Faden 1990). Excitatory amino acids such as glutamate and aspartate have been implicated in a number of pathologic conditions of the CNS including ischemia, epilepsy, neurodegenerative diseases, and more recently trauma (Meldrum 1985; Rothman and Olney 1986; Faden et al. 1989). The hypothesis that excitatory amino acids may be involved in the development of irreversible tissue injury following neurotrauma is based on observations similar to those made for opioid peptides: Excitatory amino acids accumulate in vulnerable regions following traumatic injury (Faden et al. 1989), and the detrimental effects of trauma on neurological outcome can be attenuated by excitatory amino acid antagonists (Faden and Simon 1988; Hayes et al. 1988; McIntosh et al. 1989 c; Faden et al. 1989). Again, a particular subgroup of receptors has been implicated, namely the NMDA receptors.

Membrane Breakdown and By-products

The breakdown of membranes occurs as a result of activation of phospholipases that degrade the phospholipids making up the membrane. In an elegant study, Wei et al. (1982) have directly shown that phospholipase C is activated following fluid percussion brain injury in cats. As a result, a number of membrane by-products may be generated including fatty acids, eicosanoids, and oxygen free radicals. Many of these by-products have been measured following neurotrauma, and their concentrations have been shown to increase (Ellis et al. 1981; Demediuk et al. 1985; Hsu et al. 1985; Kontos et al. 1985; Saunders et al. 1987; DeWitt et al. 1988). Various groups have also successfully attenuated traumatically induced neurological deficits by treating for these proposed injury factors with specifically targeted protective agents (Saunders et al. 1987; Hall et al. 1987; Faden et al. 1988 a).

Ion Changes

Most studies of ion changes in neurotrauma have been restricted to Na^+, K^+, and Ca^{2+} (Young 1985 a, b). Furthermore, it is only recently that these studies have been applied to fluid percussion brain injury. Studies of both spinal cord and brain injury have demonstrated increases in the total tissue Na^+ and decreases in the total tissue K^+ levels at the injury site (Lewin et al. 1974;

Lemke et al. 1987; Vink et al. 1987 a). These changes in the monovalent ions were generally thought to be representative of edema development. As such, is was not clear whether the changes were an epiphenomenon of injury or a factor contributing directly to the development of tissue damage. Recent studies have shown that neither the monovalent ion changes nor the edema development after spinal cord trauma are correlated with injury severity (Lemke and Faden 1990). It is also relevant that during the acute phase of moderate fluid percussion brain injury, while the ion changes are similar to those observed in spinal cord trauma, no significant edema developed (Vink et al. 1987 a). The observations that the extracellular Ca^{2+} concentration decreases following trauma (Stokes et al. 1983) and that the total tissue Ca^{2+} concentration increases following spinal cord injury (Young 1985 b) led to the proposal that Ca^{2+} rushes into the intracellular space following trauma, thus activating what has become known as the "calcium cascade" (Siesjo and Wieloch 1985). No measurements have appeared to date demonstrating that an increase in intracellular free Ca^{2+} does in fact occur after trauma, although the circumstantial evidence is convincing, particularly in an energy-depleted model such as spinal cord trauma (see Bioenergetics section). Even if measurements of free intracellular Ca^{2+} were obtained, it is unclear whether changes would be observed given that the cell has available numerous Ca^{2+} buffering mechanisms that could be activated (Young 1985 b). No studies have documented Ca^{2+} homeostasis following brain trauma yet.

Mg^{2+} in Neurotrauma

Brain

Until the late 1980s, it was commonly believed that the intracellular free Mg^{2+} concentration was well buffered and did not undergo significant changes in vivo. This was due in part to the fact that reliable methods for intracellular Mg^{2+} determinations were not available, and a regulatory role for Mg^{2+}, while implicated (Grubbs and Maguire 1987), could never be demonstrated. The first observation that the intracellular free Mg^{2+} concentration can undergo significant changes in vivo was made in neurotrauma by Vink et al. (1987 a). Using noninvasive MRS techniques, it was found that the free Mg^{2+} concentration decreased by 60%–70% after traumatic brain injury (Fig. 1). While this method of determination is considered by some to be somewhat controversial (see Gupta et al. 1983), under appropriate conditions, it is highly reproducible and gives comparable results to other methods (Veloso et al. 1973; Ebel and Gunther 1980; Rotevatn et al. 1989). As a discussion of the method and limitations is beyond the scope of the current review, the reader is referred to some excellent reviews (Gupta et al. 1983, 1984) and a number of original articles which critically assess the method under a variety of conditions (Kushmerick et al. 1986; Vink et al. 1988 b; Adam et al. 1988).

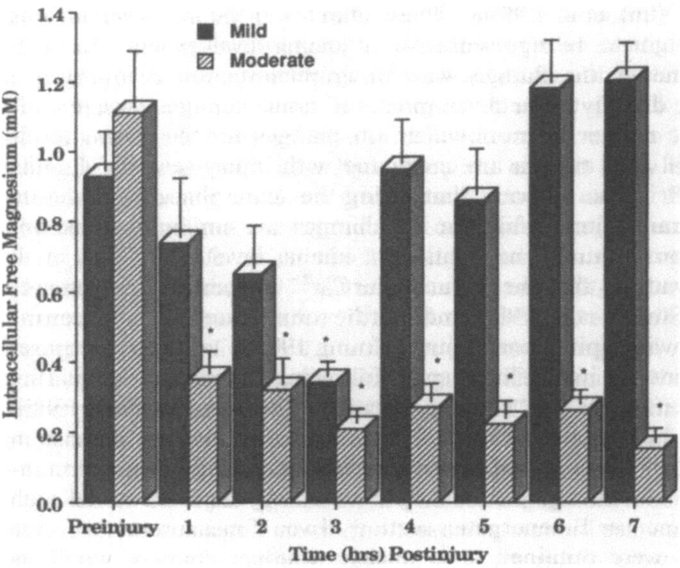

Fig. 1. Alterations in intracellular free Mg^{2+} concentration with time following mild and moderate severity fluid percussion brain injury in the rat. Free Mg^{2+} concentration was determined from phosphorus magnetic resonance spectra using the chemical shift of the β-phosphorus of ATP relative to the α-phosphorus of ATP as described in detail by Gupta et al. (1983). Adapted from the data of Vink et al. (1988 b). *$P<0.05$ vs. preinjury values

The observation in traumatic brain injury that the intracellular free Mg^{2+} level falls by up to 70% was supported by the observation that these declines were accompanied by total tissue Mg^{2+} concentration decreases of up to 10%–20% (Vink et al. 1987 a, 1988 b). While investigating hepatocyte function, Corkey et al. (1986) postulated that any observed change in total Mg^{2+} concentration may be associated with large shifts in the free Mg^{2+} concentration because changes in the bound Mg^{2+} pool would be "amplified" in the free pool which represents such a small fraction of the total tissue concentration. The declines in total and free Mg^{2+} levels in the brain provided support for this hypothesis, showing that small decreases in total concentration are associated with large declines in free concentration as would be predicted from their relative distribution in CNS tissue (i.e., the free Mg^{2+} concentration in the CNS is 8%–10% of the total tissue concentration). These changes are highly localized to the injured tissue and are not apparent in control tissue from other brain regions (Fig. 2). Furthermore, the extent of the decline, at least in the mild and moderate degrees of injury achieved by fluid percussion brain injury, was linearly correlated to the severity of injury (Vink et al. 1988 b). Only following mild injury was there any recovery of the free Mg^{2+} level in the acute phase (Fig. 1); more severe injury was associated with a decline in the free Mg^{2+} level that remained depressed for up to 12 h posttrauma. No studies have examined later time points. The total Mg^{2+} concentration did not show any

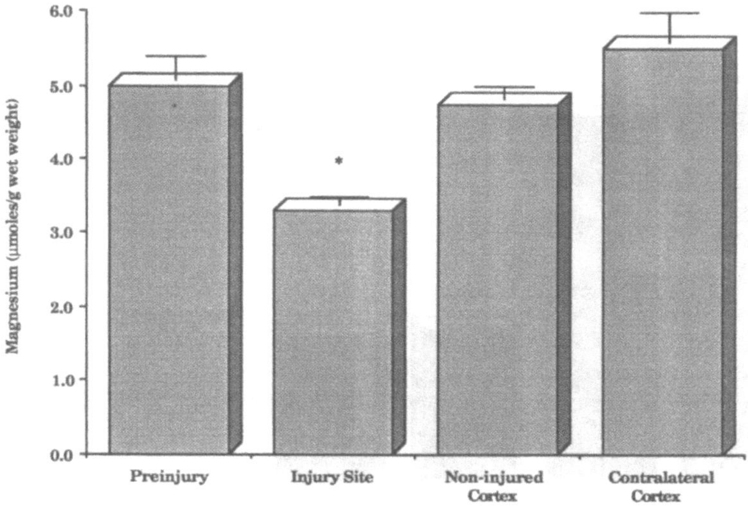

Fig. 2. Regional total tissue Mg^{2+} concentrations at 4 h after fluid percussion brain injury in the rat as determined by atomic absorption spectrophotometry. $*P<0.05$ vs. preinjury values

significant change immediately following trauma. However, by 4 h the decline was significant and remained so for at least 24 h (Fig. 3) and up to 3 days following the injury. To demonstrate further a role of Mg^{2+} in irreversible tissue injury, these authors (Vink et al. 1988 b) demonstrated that the decline in the free Mg^{2+} level after injury could be attenuated by prophylactically infusing $MgSO_4$. There was also an associated improvement in neurological outcome under these conditions. Conversely, decreasing the intracellular brain Mg^{2+} content by dietary means resulted in a significantly worsened neurological outcome after traumatic brain injury when compared with pair fed controls (McIntosh et al. 1988 a).

Improved neurological outcome after traumatic brain injury could also be demonstrated when animals were treated with intravenous $MgSO_4$ 30 min after the traumatic event (McIntosh et al. 1989 c). This effect could not be demonstrated with Mg^{2+}-ATP therapy, which has been effective in the treatment of shock and non-CNS ischemia (Chaudry et al. 1986). Since loss of autoregulation is a feature of CNS injury (Cooper 1985), this negative result was presumably caused by the profound hypotension induced by ATP infusion, which when superimposed on the existing posttraumatic hypotension would exacerbate any hypoxic/ischemic damage that the brain injury itself may have caused. In a bioenergetic analysis of their data, Vink et al. (1988 a) demonstrated that the free Mg^{2+} concentration was linearly correlated to the cytosolic phosphorylation potential (Fig. 4), presumably through actions at the creatine kinase level. This finding is particularly relevant with respect not only to the accurate estimation of phosphorylation potentials but also with

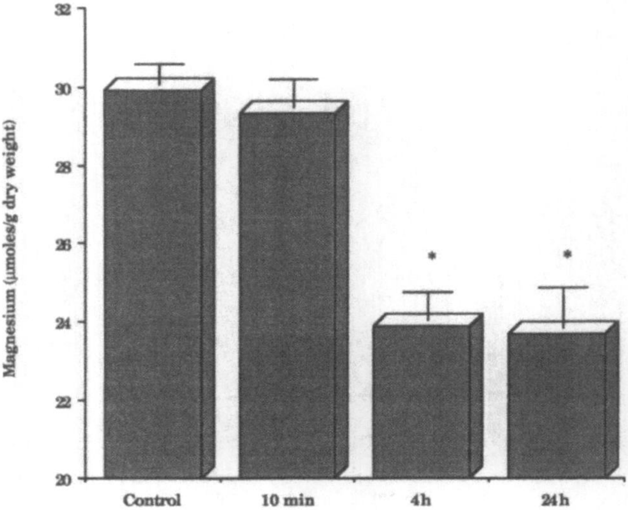

Fig. 3. Time course for total tissue Mg^{2+} concentration changes at the brain injury site following fluid percussion brain injury in the rat. $^{*}P<0.05$ vs. control (noninjured) values

Fig. 4. Relationship between brain intracellular free Mg^{2+} concentration and cytosolic phosphorylation potential prior to and following fluid percussion brain injury in rats. Adapted from data of Vink et al. (1988 a)

respect to creatine kinase dependent determinations of intracellular pH (Siesjo 1978). Clearly, without a knowledge of the Mg^{2+} concentration, these calculations cannot be performed.

Spinal Cord

Following the observation in brain injury that both total and free Mg^{2+} levels decline following trauma, Lemke et al. (1987) demonstrated that the total Mg^{2+} concentration declines following traumatic spinal cord injury. Again, the decline in total Mg^{2+} concentration was of the order of 10%–20% and was localized to the injury site (Fig. 5). Furthermore, as in brain injury, the extent of decline was related to the severity of injury (Fig. 6). To demonstrate that the declines in total Mg^{2+} concentration were not an epiphenomenon of gross tissue necrosis and were in fact representative of abnormal Mg^{2+} homeostasis in surviving tissue, MRS methods were applied to traumatic spinal cord injury to measure the free Mg^{2+} concentration (Vink et al. 1989 b). The difficulty was that ATP depletion was a feature of spinal cord trauma not apparent in brain injury. After titrating the severity of injury down to a level at which ATP was preserved for at least 2 h posttrauma yet neurological deficit was still observed at 24 h, it was demonstrated that the free Mg^{2+} concentration indeed decreased from 0.8 mM to approximately 0.3 mM by 2 h after injury (Vink et al. 1989 b). This degree of free Mg^{2+} concentration decline was similar to that observed following traumatic brain injury. Thus, the decline in

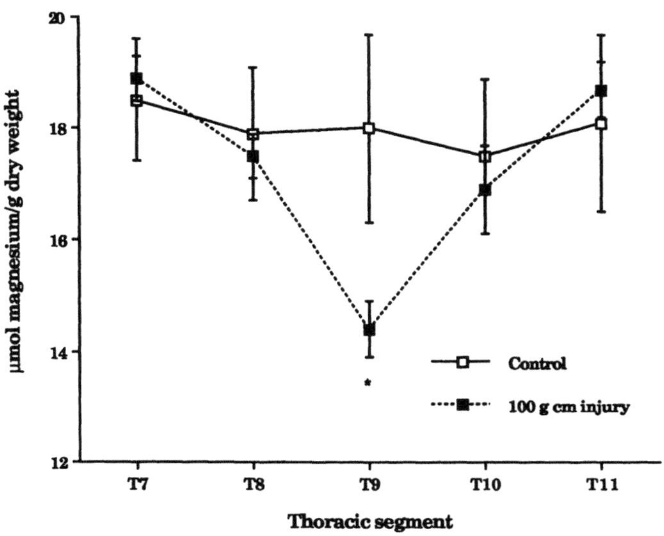

Fig. 5. Regional total tissue Mg^{2+} concentration in spinal cord at 4 h following traumatic spinal cord injury in the rat as determined by atomic absorption spectrophotometry. Weight drop injury was induced at thoracic segment T9. *$P<0.05$ vs. control (noninjured)

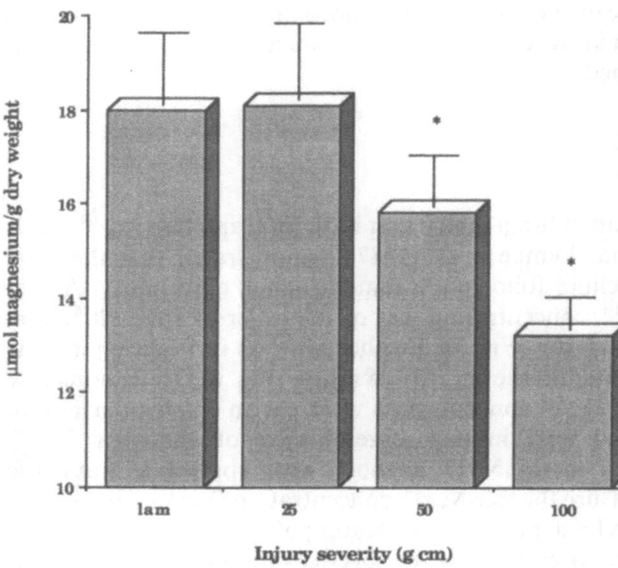

Fig. 6. Total tissue Mg^{2+} concentration (segment T9) versus injury severity following weight drop traumatic spinal cord injury in the rat. *$P<0.05$ vs. noninjured laminectomy controls

free Mg^{2+} concentration after trauma could be inferred from decreases in the total Mg^{2+} concentration. As previously observed in brain injury, the decline in total Mg^{2+} concentration was maintained for several days following the traumatic event.

Relationship to Secondary Injury

Although the role of Mg^{2+} in the CNS has been discussed previously, the relationship between the cation and the various proposed secondary injury factors has not been highlighted. Mg^{2+} plays potentially a role in each of the injury factors discussed and may exert its effects on a number of other events presumably of importance in the recovery of CNS tissue from injury.

Mg^{2+} is essential for a number of metabolic pathways, including glycolysis, Krebs cycle activity, and oxidative phosphorylation (Aikawa 1981; Heaton 1980; Garfinkel and Garfinkel 1985). As such, following neurotrauma, any changes in the Mg^{2+} concentration may be an important regulatory factor in glycolysis, substrate level phosphorylation, and oxidative phosphorylation. Alterations in bioenergetic state following trauma may then be related to alterations in Mg^{2+} concentration. Indeed, a linear relationship between the cytosolic phosphorylation potential and free Mg^{2+} concentration has already been demonstrated (Vink et al. 1988 a). Whether these changes are dependent on Mg^{2+} regulation of glycolysis or effects on mitochondrial respiration in vivo are unknown at present. Alterations in lactate production and glycolytic rate (Rawe et al. 1981) have been observed in neurotrauma. Other pathways dependent upon optimum Mg^{2+} concentration include RNA aggregation, DNA synthesis, and protein synthesis (Rubin 1976; Gunther et al. 1984). Moreover, the studies by Rubin (1976) demonstrated that when intracellular free Mg^{2+} concentration declines to levels approaching 0.25 m*M*, there is a logarithmic decline in these anabolic functions. The levels observed in neurotrauma approach this threshold. Rubin (1976) also hypothesizes that the Mg^{2+} ion concentrations are responsible for coordinate control of metabolism.

Maintenance of plasma membrane integrity is dependent upon Mg^{2+} (Ebel and Gunther 1980; Bara and Guiet-Bara 1984). Decreased membrane Mg^{2+} concentrations increase membrane fluidity and permeability to ions. One can speculate that the activities of membrane-bound enzymes such as the Na^{+}/K^{+}-ATPase may be influenced not only by Mg^{2+}-dependent ATPase activity but also by changes in membrane fluidity. It is possible that edema formation following spinal trauma may therefore be related to energy availability and to fluidity-dependent alterations in enzyme activity. Such a complex interrelationship between energy availability, membrane fluidity, enzyme activity, and edema formation would render simple correlations between any of these factors and severity of injury improbable. A relationship between the Mg^{2+} and K^{+} contents has been previously demonstrated in a variety of tissues (see Bara and Guiet-Bara 1984; Adam et al. 1988). Following traumatic CNS injury, any decline in Mg^{2+} level was also associated with a decline in K^{+} concentration. However, the kinetics of decline and recovery of K^{+} and Mg^{2+} did not correlate, perhaps illustrating that complex interactions do exist which complicate these simple correlations. Also applicable is the observation that Mg^{2+} deficiency increases fatty acid and prostanoid synthesis (Nigam et al. 1986), both of which have been implicated as secondary injury factors. With respect to recovery from membrane damage, the reacylation step of fatty acids in the resynthesis of phospholipids is also Mg^{2+} dependent, the substrate for the reaction being Mg-ATP.

Mg^{2+} plays a critical role in the maintenance of vascular tone (B. T. Altura and B. M. Altura 1982). Increasing the Mg^{2+} concentration results in dilation of cerebral arterioles. Conversely, decreasing the Mg^{2+} content of cerebral blood vessels induces Ca^{2+} mediated vasospastic responses. Moreover, under Mg^{2+}-deficient conditions, cerebral blood vessels are hyperreactive to prostanoids and leukotrienes (B. M. Altura and B. T. Altura 1985b). Focal reductions in blood flow have been documented in both brain and spinal cord injury. Reduced intracellular and extracellular Mg^{2+} concentrations may therefore exacerbate posttraumatic vasospasm and result in more severe secondary hypoxic insult.

The hypothesis that excitatory amino acids participate in the development of secondary tissue damage may also be subject to Mg^{2+} regulation. Extracellular Mg^{2+} ions exert a gating effect on excitatory amino acid receptor channels in both the brain and spinal cord (Mayer et al. 1984; Nowak et al. 1984).

The voltage-dependent blockage of the NMDA ion channels by Mg^{2+} may determine their involvement in the injury process. The release of excitatory amino acids is also thought to be regulated by Mg^{2+} (Rothman and Olney 1986). Recently, Choi (1987) has shown that Mg^{2+} can block glutamate-mediated neurotoxicity in cortical cell cultures. While such neurotoxicity has been shown to be Ca^{2+} dependent through actions at the NMDA channel, Mg^{2+} regulates all known Ca^{2+} transport mechanisms (B. M. Altura and B. T. Altura 1985b). In fact, Iseri and French (1984) has termed Mg^{2+} "nature's physiologic calcium antagonist." Although the relationship between total, intracellular, and extracellular Mg^{2+} levels has not been established following trauma, the decrease in total Mg^{2+} concentration following trauma indicates that the cation is leaving the injury site, and presumably all Mg^{2+} pools are similarly affected. Also of interest is the relationship between intracellular free Mg^{2+} concentration and NMDA receptor activity, although this is yet to be investigated (but see Johnson and Ascher, this volume).

Finally, Mg^{2+} has also been linked to opiate receptor function. At physiologic concentrations it can directly affect endogenous opioid receptor binding (Sadee et al. 1982). Given the recent evidence suggesting that opioid activity involves both opioid receptor and nonopioid receptor mechanisms (Caudle and Isaac 1988), including perhaps the excitatory amino acid receptors (Bakshi and Faden 1990; Fàden 1990), it is interesting to note tha Mg^{2+} plays a critical role in both processes.

Therapies in Neurotrauma

A number of therapies have been tried in neurotrauma (see Faden 1985). Typically, they have been applied based on their presumed protective effects against a certain predetermined injury factor. This approach has met with some success, although its degree has been limited in view of the multitude of proposed injury factors. With the aid of MRS, some of the therapies that have been shown to be beneficial to outcome can be examined with respect to their effects on the intracellular free Mg^{2+} concentration. Three groups have been examined to date: opiate antagonists, thyrotropin-releasing hormone analogues, and excitatory amino acid antagonists.

Faden and colleagues (1981, 1988c, Faden 1985) have demonstrated the efficacy of opiate antagonists in neurotrauma. In both head injury and spinal cord injury models, opiate antagonists including naloxone and nalmefene have been shown effective in improving outcome when administered up to 30 min after the traumatic insult (Young et al. 1981; Hayes et al. 1983; Faden et al. 1988c). MRS studies of the $\varkappa$ active antagonist nalmefene have demonstrated that it significantly increases intracellular free Mg^{2+} levels within 1 h after administration (Vink et al. 1990). The Mg^{2+} concentration remains elevated at approximately 80% of preinjury values for at least 6 h after administration. No studies have examined later time points. The increased free Mg^{2+} concentration is reflected in an improved cytosolic phosphorylation potential after treatment, although this improvement was not as pronounced as that observed for the free Mg^{2+} concentration. Nonetheless, improvement in neurological outcome was highly significant, suggesting a link between free Mg^{2+} concentration and recoverability.

Thyrotropin-releasing hormone analogues have been shown to be beneficial in shock and traumatic brain injury (Faden 1985). McIntosh et al. (1988 b) demonstrated that administration of the analogue CG 3703 30 min after moderate fluid percussion head injury significantly improved cardiovascular variables, cellular bioenergetic status, and neurological motor function, although the mechanisms for this improved status are unknown. Recently, it has been demonstrated that administration of this analogue after traumatic head injury increases both intracellular free and total Mg^{2+} concentrations when compared with vehicle-treated controls (Vink et al. 1988 c).

Finally, two excitatory amino acid antagonists have been examined with respect to their effects on Mg^{2+} homeostasis following fluid percussion induced brain injury. Dextrorphan was shown to increase concentrations of intracellular free Mg^{2+} to up to 90% of preinjury values, whereas the saline-treated controls had suppressed free Mg^{2+} concentrations of the order of 30% of preinjury values (Faden et al. 1989). Notably there was a significant improvement in neurologic recovery in dextrorphan-treated animals. The other antagonist studied to date has been MK 801, a noncompetitive antagonist widely used with some success in ischemia (Kemp et al. 1987; Olney et al. 1987). When administered 30 min after moderate fluid percussion head injury, it caused a transient increase in intracellular free Mg^{2+} concentration (McIntosh et al. 1990). The Mg^{2+} level increased to approximately 70% of preinjury values but returned to saline-treated values within 2 h after treatment. The authors rationalized the short and transient effects of MK 801 in terms of its relatively short plasma half-life (Wong et al. 1986) and suggested that more effective treatment may be obtained by using continuous infusion rather than a single bolus injection. In relation to the present discussion, it is of interest to note that the beneficial effects of this drug on the neurological outcome in this particular study were also very modest (T. K. McIntosh, unpublished results). In contrast, when administered following spinal cord trauma, MK 801 resulted in a significant improvement in total Mg^{2+} concentration at the injury site and improved chronic neurological outcome (Faden et al. 1988 b).

Future Directions

It is clear from the discussion above that there is a potential role for Mg^{2+} in neurotrauma and perhaps other pathologies of CNS function. Several groups have successfully used $MgSO_4$ in treatment for ischemia, pre-eclampsia, migraine, and epilepsy (Vacanti and Ames 1984; Altura and Altura 1985 a; Robertson et al. 1986). Other investigators have reported data that strongly implicate Mg^{2+} as an important injury factor in both ischemia (Ginsburg et al. 1977; Yoshida et al. 1984) and clinical head trauma (Cavaliere et al. 1985).

Nonetheless, until recently little direct evidence of a potential role for Mg^{2+} was forthcoming. The use of MRS technology provides an opportunity by which effects of injury and treatment on the free Mg^{2+} status can be monitored directly. Reports from other laboratories are now appearing demonstrating significant alterations in the free Mg^{2+} concentration, changes that have the potential to exert considerable metabolic control over cell functions (Adam et al. 1988; Borchgrevink et al. 1989; Brooks and Bachelard 1989). It remains to be established whether these changes are capable of inducing a secondary injury cascade in themselves or act in concert with other factors to produce irreversible tissue damage. Cetainly the concept of metabolic control by Ca^{2+}/Mg^{2+} ratios is not new (Bygrave 1978; Cittadini et al. 1982). It is the characterization of this relationship that will ultimately establish the role of these critically important cations in the pathophysiology of irreversible tissue injury. This cation balance may provide the key for appropriate therapeutic interventions.

Acknowledgement. R. V. is a Queen Elizabeth II Research Fellow and is supported by the Australian National Research Fellowships Advisory Committee.

References

Adam WR, Craik DJ, Hall JG, Kneen MM, Wellard RM (1988) Problems in the assessment of magnesium depletion in the rat by in vivo 31P NMR. Magn Reson Med 7:300–310
Aikawa JK (1981) Magnesium: it's biologic significance. CRC, Boca Raton
Allen AR (1911) Surgery of experimental lesion equivalent to crush injury of fracture dislocation of spinal column. JAMA 57:878–880
Altura BM, Altura BT (1985a) New perspectives on the role of magnesium in the pathophysiology of the cardiovascular system. I: clinical aspects. Magnesium 4:226–244
Altura BM, Altura BT (1985b) New perspectives on the role of magnesium in the pathophysiology of the cardiovascular system. II: experimental aspects. Magnesium 4:245–271
Altura BM, Altura BT, Gebrewold A, Ising H, Gunther T (1984) Magnesium deficient diets and microcirculatory changes in situ. Science 223:1315–1317
Altura BT, Altura BM (1982) The role of magnesium in etiology of strokes and cerebrospasm. Magnesium 1:277–291
Bakshi R, Faden AI (1990) Competitive and non-competitive NMDA antagonists limit dynorphinA-induced rat hindlimb paralysis. Brain Res 507:1–5
Bara M, Guiet-Bara A (1984) Potassium, magnesium and membranes. Magnesium 3:212–225
Borchgrevink PC, Bergan AS, Bakoy OE, Jynge P (1989) Magnesium and reperfusion of ischemic rat heart as assessed by 31P-NMR. Am J Physiol 256:H195–H204
Brooks KJ, Bachelard HS (1989) Changes in intracellular free magnesium during hypoglycemia and hypoxia in cerebral tissue as calculated from 31P-nuclear magnetic resonance spectra. J Neurochem 53:331–334
Bygrave FL (1978) Mitochondria and the control of intracellular calcium. Biol Rev 53:43–79
Caudle RM, Isaac L (1988) A novel interaction between dynorphin (1–13) and an n-methyl-D-aspartate site. Brain Res 443:329–332
Cavaliere F, Sciarra M, Crea MA, Rossi M, Proietti R (1985) Variazioni del magnesio sierico ed urinario in pazienti traumatizzati cranici. Recenti Prog Med 76:563–566
Chaudry IH, Clemens MG, Baue AE (1986) The role of ATP-magnesium in ischemia and shock. Magnesium 5:211–220

Choi D (1987) Ionic dependence of glutamate neurotoxicity. J Neurosci 7:369–379
Cittadini A, Bossi D, Wolf FI, Dani AM (1982) The role of the intracellular Ca/Mg ratio in bioenergetic reactions. In: Anghileri LJ, Tuffet-Anghileri AM (eds) The role of calcium in biological systems, vol 1. CRC, Boca Raton, pp 189–200
Cooper PR (1985) Delayed brain injury: secondary insults. In: Becker DP, Povlishock JT (eds) Central nervous system trauma status report 1985. National Institute of Health, Bethesda, pp 217–228
Corkey BE, Duszynski J, Rich TL, Matschinsky B, Williamson JR (1986) Regulation of free and bound magnesium in rat hepatocytes and isolated mitochondria. J Biol Chem 261:2567–2574
Demediuk P, Saunders RD, Clendenon NR, Means ED, Anderson DK, Horrocks LA (1985) Changes in lipid metabolism in traumatized spinal cord. Prog Brain Res 63:1–16
DeSalles AF, Kontos HA, Becker DP, Yang MS, Ward JD, Moulton R, Gruemer HD, Lutz H, Maset AL, Jenkins L, Marmarou A, Muizelaar P (1986) Prognostic significance of ventricular CSF lactic acidosis in severe head injury. J Neurosurg 65:615–624
DeWitt DS, Kong DL, Lyeth BG, Jenkins LW, Hayes RL, Wooten ED, Prough DS (1988) Experimental traumatic brain injury elevates brain prostaglandin E_2 and thromboxane B_2 levels in rats. J Neurotrauma 4:303–313
Dixon CE, Lyeth BG, Povlishock JT (1987) A fluid percussion model of experimental brain injury in the rat. J Neurosurg 67:110–119
Ebel H, Gunther T (1980) Magnesium metabolism: a review. J Clin Chem Clin Biochem 18:257–270
Ellis EF, Wright KF, Wei EP, Kontos HA (1981) Cyclooxygenase products of arachidonic acid metabolism in cat cerebral cortex after experimental concussive brain injury. J Neurochem 37:892–896
Enevoldsen EM, Cold G, Jensen FT (1976) Dynamic changes in regional CBF, intraventricular pressure, CSF, pH and lactate levels during the acute phase of head injury. J Neurosurg 44:191–214
Faden AI (1985) Pharmacologic therapy in acute spinal cord injury: experimental strategies and future directions. In: Becker DP, Povlishock JT (eds) Central nervous system trauma status report 1985. National Institute of Health, Bethesda, pp 481–485
Faden AI (1990) Opioid and non-opioid mechanisms may contribute to dynorphin's pathophysiologic actions in spinal cord injury. Ann Neurol 27:67–74
Faden AI, Simon RP (1988) A potential role for excitotoxins in the pathophysiology of spinal cord injury. Ann Neurol 23:623–626
Faden AI, Jacobs TP, Holaday JW (1981) Opiate antagonists improve neurologic recovery after spinal injury. Science 211:493–494
Faden AI, Molineaux CJ, Rosenberger JG, Jacobs TP, Cox BM (1985) Endogenous opioid immunoreactivity in rat spinal cord following traumatic injury. Ann Neurol 17:386–390
Faden AI, Lemke M, Demediuk P (1988a) Effects of BW755C, a mixed cyclo-oxygenase-lipoxygenase inhibitor, following traumatic spinal cord injury in rats. Brain Res 463:63–68
Faden AI, Lemke M, Simon RP, Noble LJ (1988b) N-methyl-D-aspartate antagonist MK801 improves outcome following traumatic spinal cord injury in rats: behavioural, anatomic, and neurochemical studies. J Neurotrauma 5:33–45
Faden AI, Sacksen I, Noble LJ (1988c) Opiate receptor antagonist nalmefene improves neurological recovery after traumatic spinal cord injury in rats through a central mechanism. J Pharmacol Exp Ther 245:742–748
Faden AI, Demediuk P, Panter S, Vink R (1989) Excitatory amino acids, n-methyl-D-aspartate receptors and traumatic brain injury. Science 244:798–800
Frankowski RF, Annegers JF, Whitman S (1985) Epidemiological and descriptive studies pt 1: the descriptive epidemiology of head trauma in the United States. In: Becker DP, Povlishock JT (eds) Central nervous system trauma status report 1985. National Institute of Health, Bethesda, pp 33–43
Garfinkel L, Garfinkel D (1985) Magnesium regulation of the glycolytic pathway and the enzymes involved. Magnesium 14:60–72

Gennarelli TA, Thibault LE (1985) Biological models of head injury. In: Becker DP, Povlishock JT (eds) Central nervous system trauma status report 1985. National Institute of Health, Bethesda, pp 391–404
Ginsburg MD, Mela L, Wrobel-Kuhl K, Reivich M (1977) Mitochondrial metabolism following bilateral cerebral ischemia in the gerbil. Ann Neurol 1:519–527
Grubbs RD, Maguire ME (1987) Magnesium as a regulatory cation: criteria and evaluation. Magnesium 6:113–127
Gulati SC, Sood SC, Bali IM, Kak VK (1980) Cerebral metabolism following brain injury. Acta Neurochir (Wien) 53:39–51
Gunther T, Vormann J, Forster R (1984) Functional compartmentation of intracellular magnesium. Magnesium-Bull 6:77–81
Gupta RK, Gupta P, Yushok WD, Rose ZB (1983) On the noninvasive measurement of intracellular free magnesium by 31P NMR spectroscopy. Physiol Chem Phys Med NMR 15:265–280
Gupta RK, Gupta P, Moore RD (1984) NMR studies of intracellular metal ions in intact cells and tissue. Annu Rev Biophys Bioeng 13:221–246
Hall ED, McCall JM, Chase RL, Yonkers PA, Braughler JM (1987) A non-glucocorticoid steroid analog af methylprednisolone duplicates its high dose pharmacology in models of central nervous system trauma and neuronal membrane damage. J Pharmacol Exp Ther 242:137–142
Hayes RL, Galinet BJ, Kulkarne P (1983) Effects of naloxone on systemic and cerebral responses to experimental concussive brain injury in cats. J Neurosurg 58:720–728
Hayes RL, Jenkins LW, Lyeth BG, Balster RL, Robinson SE, Clifton GL, Stubbins JF, Young HF (1988) Pretreatment with phencyclidine, an n-methyl-D-aspartate antagonist, attenuates long term behavioral deficits in the rat produced by traumatic brain injury. J Neurotrauma 5:259–274
Heaton FW (1980) Magnesium in intermediary metabolism. In: Cantin M, Seelig MS (eds) Magnesium in health and disease. Spectrum, New York, pp 43–55
Hillered L, Siesjo BK, Arfors K-E (1984) Mitochondrial response to transient forebrain ischemia and recirculation in the rat. J Cereb Blood Flow Metab 4:438–446
Hsu CY, Halushka PV, Hogan EL, Banik NL, Lee WA, Perot LP (1985) Alteration of thromboxane and prostacyclin levels in experimental spinal cord injury. Neurology 35:1003–1009
Iseri LT, French JH (1984) Magnesium: nature's physiologic calcium antagonist. Am Heart J 108:188–193
Kemp JA, Foster AC, Wong EHF (1987) Noncompetitive antagonists of excitatory amino acid receptors. Trends Neurosci 10:294–298
Kontos HA, Wei EP, Ellis EF et al. (1985) Appearance of superoxide anion radical in cerebral extracellular space during increased prostaglandin synthesis in cats. Circ Res 57:142–151
Kushmerick MJ, Dillon PF, Meyer RA, Brown TR, Krisanda JM, Sweeney HL (1986) 31P NMR spectroscopy, chemical analysis, and free Mg^{2+} of rabbit bladder and uterine smooth muscle. J Biol Chem 261:14420–14429
Kwo S, Young W, DeCrescito V (1989) Spinal cord sodium, potassium, calcium and water concentration changes in rats after graded contusion injury. J Neurotrauma 6:13–24
Lemke M, Demediuk P, McIntosh TK, Vink R, Faden AI (1987) Alterations in tissue Mg^{++}, Na^{+} and spinal cord edema following impact trauma in rats. Biochem Biophys Res Commun 149:594–599
Lemke M, Faden AI (1990) Edema development and ion changes to rat spinal cord after impact trauma: injury dose response studies. J Neurotrauma 7:41–54
Lewin MG, Hansebout RH, Pappius HM (1974) Chemical characteristics of traumatic spinal cord edema in cats. J Neurosurg 40:65–75
Nayer ML, Westbrook GL, Cuthrie PB (1984) Voltage dependent block by Mg^{2+} of NMDA receptors in spinal cord neurones. Nature 309:261–263
McIntosh TK, Vink R (1989) Biochemical and pathophysiologic mechanisms in experimental mild to moderate traumatic brain injury. In: Doran J Jr, Anderson TE, Cole TM (eds)

Contemporary issues in neurological surgery, vol 1; mild to moderate head injury. Blackwell, Boston, pp 35–46
McIntosh TK, Faden AI, Bendall MR, Vink R (1987) Traumatic brain injury in the rat: alterations in brain lactate and pH as characterized by 1H and 31P NMR. J Neurochem 49:1530–1540
McIntosh TK, Faden AI, Yamakami I, Vink R (1988a) Magnesium deficiency exacerbates and pretreatment improves outcome following traumatic brain injury in rats: 31P magnetic resonance spectroscopy and behavioral studies. J Neurotrauma 5:17–31
McIntosh TK, Vink R, Faden AI (1988b) An analog of thyrotropin-releasing hormone improves outcome after brain injury: 31P NMR studies. Am J Physiol 254:R785–R792
McIntosh TK, Soares HL, Hayes RL, Simon RP (1989a) The *n*-methyl-D-aspartate receptor antagonist MK 801 prevents edema and restores magnesium homeostasis following traumatic brain injury in rats. In: Leahman J (ed) Recent advances in excitatory amino acid research. Liss, New York, pp 653–656
McIntosh TK, Vink R, Noble LJ, Yamakami I, Fernyak SE, Soares H, Faden AI (1989b) Traumatic brain injury in the rat: characterization of a lateral fluid percussion injury model. Neuroscience 28:233–244
McIntosh TK, Vink R, Soares H, Hayes R, Simon R (1989c) Effects of the *N*-methyl-D-aspartate receptor blocker MR 801 on neurological function after experimental brain injury. J Neurotrauma 6:247–259
McIntosh TK, Vink R, Yamakami I, Faden AI (1989d) Magnesium protects against neurological deficit after brain injury. Brain Res 482:252–260
McIntosh TK, Vink R, Soares H, Simon RP (1990) Effect of noncompetitive blockade of *N*-methyl-D-aspartate receptors on the neurochemical sequelae of experimental brain injury. J Neurochem (in press)
Meldrum B (1985) Possible therapeutic applications of antagonists of excitatory amino acid neurotransmitters. Clin Sci 68:113–122
Nigam R, Averdunk Y, Gunther T (1986) Alteration in prostanoid metabolism in rats with magnesium deficiency. Prostaglandins Leukot Med 23:1–10
Nilsson B, Ponten U (1977) Experimental head injury in the rat, pt 2: regional brain energy metabolism in concussive trauma. J Neurosurg 47:252–261
Noble LJ, Wrathall JR (1989) Correlative analysis of lesion development and functional status after graded spinal cord contusive injuries in the rat. Exp Neurol 103:34–40
Nowak L, Bregestovski P, Ascher P, Herbelt A, Prochiantz A (1984) Magnesium gates glutamate activated channels in mouse central neurones. Nature 307:462–465
Olney JW, Price M, Salles K, Labruyere J, Friedrich G (1987) MK 801 powerfully protects against N-methyl-D-aspartate neurotoxicity. Eur J Pharmacol 141:357–361
Povlishock JT (1985) The morphopathologic responses to head injuries of varying severity. In: Becker DP, Povlishock JT (eds) Central nervous system trauma status report 1985. National Institute of Health, Bethesda, pp 443–452
Rawe SE, Lee WA, Perot PL (1981) Spinal cord glucose utilization after experimental spinal cord injury. Neurosurgery 9:40–47
Rivlin AS, Tator CH (1978) Effect of duration of acute spinal cord compression in a new acute cord injury model in the rat. Surg Neurol 10:39–43
Robertson CS, Foltz R, Grossman RG, Goodman JC (1986) Protection against experimental ischemic spinal cord injury. J Neurosurg 64:633–642
Rosner MJ, Becker DP (1984) Experimental brain injury: successful therapy with the weak base, tromethamine. J Neurosurg 60:961–971
Rotevatn S, Murphy E, Levy LA, Raju B, Lieberman M, London RE (1989) Cytosolic free magnesium concentration in cultured chick heart cells. Am J Physiol 257:C141–C146
Rothman SM, Olney JW (1986) Glutamate and the pathophysiology of hypoxic-ischemic brain damage. Ann Neurol 19:105–111
Rubin H (1976) Magnesium deprivation reproduces the coordinate effects of serum removal or cortisol addition on transport and metabolism in chick embryo fibroblasts. J Cell Physiol 89:613–626

Sadee W, Pfeiffer A, Herz A (1982) Opiate receptors: multiple effects of metal ions. J Neurochem 39:659–667
Sandler AN, Tator CH (1976) Review of the effects of spinal cord trauma on vessels and blood flow in the spinal cord. J Neurosurg 45:638–646
Saunders RD, Dugan LL, Demediuk P, Means ED, Horrocks LA, Anderson DK (1987) Effects of methylprednisolone and the combination of a-tocopherol and selenium on arachidonic acid metabolism and lipid peroxidation in traumatized spinal cord tissue. J Neurochem 49:24–31
Selecki BR, Simpson DA, Vanderfield GK, Ring I, Sewell MK (1980) The epidemiology of head injury in New South Wales. Proceedings of the 5th annual Brain Impairment Conference, Lidcombe Hospital, Syndey
Siesjo BK (1978) Brain energy metabolism. Wiley, New York
Siesjo BK, Wieloch T (1985) Brain injury: neurochemical aspects. In: Becker DP, Povlishock JT (eds) Central nervous system trauma status report 1985. National Institute of Health, Bethesda, pp 513–532
Stewart WB (1985) Edema in spinal cord injury. In: Becker DP, Povlishock JT (eds) Central nervous system trauma status report 1985. National Institute of Health, Bethesda, pp 475–479
Stokes BT, Fox P, Hollinden G (1983) Extracellular calcium activity in the injured spinal cord. Exp Neurol 80:561–572
Sullivan HG, Martinez J, Becker DP, Miller JD, Griffith R, Wist AO (1976) Fluid percussion model of mechanical brain injury in the cat. J Neurosurg 45:520–534
Tarlov IM (1972) Acute spinal cord compression paralysis. J Neurosurg 36:10–20
Tornheim PA (1985) Traumatic edema in head injury. In: Becker DP, Povlishock JT (eds) Central nervous system trauma status report 1985. National Institute of Health, Bethesda, pp 431–442
Vacanti FX, Ames A (1984) Mild hypotension and Mg^{2+} protect against irreversible damage during CNS ischemia. Stroke 15:695–698
Veloso D, Guynn RW, Oskarsson M, Veech RL (1973) The concentration of free and bound magnesium in rat tissues. J Biol Chem 248:4811–4819
Vink R, McIntosh TK, Demediuk P, Faden AI (1987a) Decrease in total and free magnesium concentration following traumatic brain injury in rats. Biochem Biophys Res Commun 149:594–599
Vink R, McIntosh TK, Weiner MW, Faden AI (1987b) Effects of traumatic brain injury on cerebral high energy phosphates and intracellular pH: a 31P magnetic resonance spectroscopy study. J Cereb Blood Flow Metab 7:563–571
Vink R, Faden AI, McIntosh TK (1988a) Changes in cellular bioenergetic state following graded traumatic brain injury in rats: determination by phosphorus-31 magnetic resonance spectroscopy. J Neurotrauma 5:105–119
Vink R, McIntosh TK, Demediuk P, Weiner MW, Faden AI (1988b) Decline in intracellular free magnesium concentration is associated with irreversible tissue injury following brain trauma. J Biol Chem 263:757–761
Vink R, McIntosh TK, Faden AI (1988c) Treatment with the thyrotropin-releasing hormone analog CG3703 restores magnesium homeostasis following traumatic brain injury in rats. Brain Res 460:184–188
Vink R, McIntosh TK, Yamakami I, Faden AI (1988d) 31P NMR characterization of graded traumatic brain injury in rats. Magn Reson Med 6:37–48
Vink R, Noble LJ, Knoblach SM, Bendall MR, Faden AI (1989a) Metabolic changes in rabbit spinal cord after trauma: magnetic resonance spectrocopy studies. Ann Neurol 25:26–31
Vink R, Yum SW, Lemke M, Demediuk P, Faden AI (1989b) Traumatic spinal cord injury in rabbits decreases intracellular free magnesium concentration as measured by 31P MRS. Brain Res 490:144–147
Vink R, McIntosh TK, Rhomhanyi R, Faden A (1990) Opiate antagonist nalmefene improves intracellular free Mg^{2+}, bioenergetic state and neurologic outcome following traumatic brain injury in rats. J Neurosci (in press)

Walker JG, Yates RR, O'Neill JJ, Yashon D (1977) Canine spinal cord energy state after experimental trauma. J Neurochem 29:929–932
Wagner KR, Tornheim PA, Eichold ME (1985) Acute changes in regional cerebral metabolite values following experimental blunt head trauma. J Neurosurg 63:88–96
Wei EP, Lamb RG, Kontos HA (1982) Increased phospholipase C activity after experimental brain injury. J Neurosurg 56:695–698
Wong EHF, Kemp JA, Priestley T (1986) The novel anticonvulsant MK 801 is a potent N-methyl-D-aspartate antagonist. Proc Natl Acad Sci USA 83:7104–7109
Yamakami I, McIntosh TK (1989) Effects of traumatic brain injury on regional cerebral blood flow in rats as measured with radiolabeled microspheres. J Cereb Blood Flow Metab 9:117–124
Yang MS, DeWitt DS, Becker DP, Hayes RL (1985) Regional brain metabolite levels following mild experimental head injury in the cat. J Neurosurg 63:617–621
Yoshida S, Busto R, Martinez E, Scheinberg P, Ginsburg MD (1985) Regional brain energy metabolism after complete versus incomplete ischemia in the rat in the absence of severe lactic acidosis. J Cereb Blood Flow Metab 5:490–501
Young W (1985 a) Blood flow, metabolic and neurophysiological mechanisms in spinal cord injury. In: Becker DP, Povlishock JT (eds) Central nervous system trauma status report 1985. National Institute of Health, Bethesda, pp 463–473
Young W (1985 b) The role of calcium in spinal cord injury. CNS Trauma 2:109–114
Young W, Flamm ES, Demopoulos HB, Tomasula JJ, DeCrescito V (1981) Naloxone ameliorates posttraumatic ischemia in experimental spinal contusion. J Neurosurg 55:209–219
Yuan X-Q, Prough DS, Smith TL, Dewitt DS (1988) The effects of traumatic brain injury on regional cerebral blood flow in rats. J Neurotrauma 5:289–301

Subject Index

A-current 42
acetylcholine-activated K^+ channel 21, 26, 30, 42
Ach (nicotinic) receptors 116
Ach-activated channels, block by local anesthetic 108
Ach-channels block, nicotinic, by Mg^{2+} 110
Allen weight drop method 126, 129
ATP binding 36–39
ATP-dependent channels 33–35
ATP-dependent K^+ channels 33–47
– heart 33, 34, 36, 38, 39, 43
– hypothalamus, glucose response neurones 35
– block by Mg^{2+} 42–44
– nerve terminals 33
– pancreatic β cells 33, 34–39, 41
– peptides action 35, 40
– role of H^+ 45, 46
– role of Na^+ 45
– role of intracellular Mg^{2+} 33, 41–45
– skeletal muscle 33, 34, 36, 37, 39, 40, 43
– channels, smooth muscle 33, 35
ATP-induced hypotension 133

BAY K-8644 99
β-adrenergic
– agonists 84
– receptors 91
brain trauma 116, 129
8-bromo cAMP 90, 92

Ca^{2+}
– ATPase 81
– blockers 23
– buffering 74
– channels (see Ca^{2+} channels)
– concentration in neurotrauma 130
– dependent K^+ permeability 33
– homeostasis in neurotrauma 131
– independent inactivation 74, 91, 93
Ca^{2+} channels
– ω-CgTx -insensitive 98
– in Barnacle muscles 99
– in chick embryos 99
– deactivation 98
– inactivation 72, 74, 75
– L-type 24, 84, 94
– low-threshold 97, 100
– phosphorylation 72
– in snail neurons 99, 103
– T-type 84
cAMP-dependent
– phosphorylation 85
– protein kinase 90
cardiac
– action potential 33, 44, 51, 91
– myocytes 71
cerebral
– blood flow (CBF) 120
– metabolic rate for O_2 120
CG3703 139
channel
– conducting unit 64, 69
– phosphorylation 41
cholinergic receptors 91
CNS trauma 129
Co^{2+} ion 99
compression model 125, 129

delayed rectifier K^+ channel 26, 51
dextrorphan 139
2-[^{14}C]DG 120
dihydropyridine(s) 84
– agonists 99
DRG neurons 98

EDHP 36
EDTA 82
EGTA 80, 98
ETH-1117 5, 7, 9–12
ETH-5214 5, 7–11, 16
excitation-contraction coupling 91
excitatory amino acids
– effects on epilepsy 130
– effects on ischemia 130
– effects on neurodegeneration 130

excitatory CNS synapses 105
extracellular Mg^{2+} 97

fluid percussion model 126, 128, 130, 132
fluorescence method 16
fluorescent intracellular probe 1
fluorinated NMR method 1
forskolin 84
free intracellular concentration (see also intracellular free Mg^{2+}) 1

G proteins 26, 40
galanin 35, 40
GH_3 cells 92
glucose response neurones in hypothalamus 35
glutamatergic synapses 105
guanine nucleotide binding (G proteins) 26, 40

Henle loop 34
high treshold Ca^{2+} channel 21
hypermagnesemia 121
– effects on white matter 122

insulin secreting cell line (insulinoma cell line) 35, 40, 43, 45
intracellular free Mg^{2+} 1
– Ca^{2+} uptake 1
– cardiovascular disease 1
– enzymatic reactions 1
– K^+ uptake 1
– in neurotrauma 131
intracellular Mg^{2+} 21, 26, 29, 112
intracellular Mg^{2+} -effects
– on adenylate cyclase 92
– on Ca^{2+} channels 75
– on phosphodiesterase 92
– on protein kinase activity 92
inwardly rectifying K^+ channels 21, 26, 29, 34, 42, 51–69
– I–V curve 54, 58, 61, 63, 68
– I–V relation/curve 54, 58, 61, 63, 68
– blockage by Mg^{2+} 51, 52, 61–69
– outward current 51–53
– substate conductance 63
– voltage-dependent blockage 51–53, 60
Ion selective microelectrodes 1–5, 9–12
– activity coefficient 4, 14
– calibration 3, 6, 10, 12, 16
– calibration curve 2, 10, 11, 13
– selectivity 3
– sylanization 5

K^+ channels 21, 33, 51, 61, 68, 69
– Ach-activated 115
– Ach-dependent 71
– ATP binding and Mg^{2+} 36–39
– ATP-dependent 71, 115
– ATP-regulated 51, 61, 68, 69
– block by Mg^{2+} 71
– Ca^{2+}-activated 51
– cardiac action potential 44
– delayed rectifier 51, 71
– inward rectification 34, 42, 51, 69, 115
– inward rectifier 71
– muscarinic receptor-operated 51, 61, 68, 69
– Na^+-activated 51, 68
K^+ concentration in neurotrauma 130, 137
K^+ conductance 33
K^+ permeability 33, 35, 44

L-glutamate receptors 105
ligand-gated K^+ channels 21, 25, 26
long-term potentiation (LTP) 105
low-threshold (LVA) channels
– in chick embryos 97
– in neoplastic lymphocytes 100
– in neurons 102
LTP (long-term potentiation) 105

membrane patches 101
Mg^{2+} block/blockage
– apparent Kd
– in chick embryos 100
– in heart muscle 100
– inwardly rectifying K^+ channels 51, 52, 61–69
– of Na^+ currents 100
Mg^{2+} concentration
– in CNS 120, 132
– in cortex 121
– in CSF 121, 127
– during cytosolic phosphorylation 133
– in fetal plasma 121
– in gray matter 121
– in hepatocyte 132
– in maternal serum 121
– in midbrain 121
– in spinal cord injury 135
– in white matter 121
Mg^{2+} -dependent inactivation 94
Mg^{2+} electrodes 3, 10
Mg^{2+} free concentration (intracellular) 1
– cardiac muscle 12
– heart 1, 7, 12, 13, 15
– homeostasis 135
– mean value 13
– skeletal muscle 14
– smooth muscle 14

– ventricle/ventricular muscle 12,13
Mg^{2+} , intracellular 112
Mg^{2+} ion
– apparent dissociation constant 106, 108, 110, 116
– in ATP-dependent reactions 127
– in CNS 127
– as CNS depressant 119, 120
– effects (see Mg^{2+} ion effects)
– entry rate 99
– exit rate 102
– on glycolysis 119, 121
– hydration shell 97
– on hypoxia 120
– internal binding site 103, 106, 107
– iontophoretic application 120, 121
– in ischemia 139
– in neurotrauma 125
– in plasma 127
– rate of entry 108
– rate of exit 108
– voltage-dependent block 115
– water complex 103
Mg^{2+} ion effects
– on anabolic functions 137
– on catalytic activity 119
– on creatine kinase 127
– on DNA synthesis 137
– on enzymes activation 119
– on glycolysis 135
– on membrane enzymes 127, 137
– on membrane fluidity 127, 137
– on mitochondrial respiration 135
– on Na^+/K^+ -ATPase 127
– on neuromuscular tone 127
– on opiate receptors 138
– on phosphofructokinase 127
– on protein synthesis 119, 127, 137
– on RNA aggregation 135
– on vascular tone 127, 137
Mg^{2+} measurement 12, 13
– MRS technique 50, 128, 131, 135, 138
$MgSO_2$
– intravenous 133
– prophylactically infused 133
MK801 139
muscarinic atrial K^+ channel 21, 26, 30, 42

Na^+ channels 21
– in squid axons 103
– TTX-sensitive 71
Na^+ concentration in neurotrauma 130
Na^+ currents
– through cardiac Ca^{2+} channels 100
– through LVA channels 97, 99, 100
– through skeletal muscle Ca^{2+} channels 100
– TTX-insensitive 98
nalmefene 138
nickel 84
nicotinic Ach-channels block by Mg^{2+} 110
nifedipine 84
NMDA-activated channels 105, 106, 110
– block by Mg^{2+} 105, 111, 113
NMDA receptors 42, 105, 110
NMR 1, 15
– method, fluorinated 1
nucleotides binding 33
– other than ADP and ATP 40

opioids
– antagonists in neurotrauma 130, 138
– in neurotrauma 129
– receptors 128, 130

pacemaker depolarization 91
patch clamp technique 72, 98, 106
phosphodiesterase activity 87
phosphorylation 41
protein
– kinase C 41
– phophorylation 93
– phosphatase activity 90

receptors
– Ach (nicotinic) 116
– β adrenergic 92
– L-glutamate 105
– NMDA 105, 110, 128, 132
– opioids 128, 130

sarcoplasmic reticulum
– Ca^{2+} channels 34
– Ca^{2+}-pump 81
single
– file diffusion 100
– ionic channels 101, 106
sinoatrial node 91
skeletal muscle 24
snail neurons 92
spinal cord trauma 126, 128, 129
squid axon 34
subconductance states 23, 63
surface charges
– screening by Mg^{2+} 92, 97, 102

tetrametryn 84
thyrotropin-releasing hormone
– in neurotrauma 138, 139
– in shock 138, 139

ventricular cells 84
VIP 36

voltage clamp 21
voltage-dependent blockage 42, 44–46
– by H^+ 45, 46
– by Mg^{2+} 41–45, 61–69
– by Na^+ 45
voltage-dependent inactivation 75, 82

voltage-gated
– Ca^{2+} channels 25, 71
– K^+ channels 26
– K^+ current 21, 26
– Na^+ channel 21, 23, 24, 115
– Na^+ current 21, 25

X chromosome 25

MIX
Papier aus verantwortungsvollen Quellen
Paper from responsible sources
FSC® C105338

If you have any concerns about our products,
you can contact us on
ProductSafety@springernature.com

In case Publisher is established outside the EU,
the EU authorized representative is:
Springer Nature Customer Service Center GmbH
Europaplatz 3, 69115 Heidelberg, Germany

Printed by Libri Plureos GmbH
in Hamburg, Germany